PUBLIC POLICY WRITING THAT MATTERS

DAVID CHRISINGER

Johns Hopkins University Press • Baltimore

Printed in the United States of America on acid-free paper
9 8 7 6 5 4 3 2 1

Johns Hopkins University Press
2715 North Charles Street
Baltimore, Maryland 21218-4363
www.press.jhu.edu

Library of Congress Cataloging-in-Publication Data

Names: Chrisinger, David, 1986– author.
Title: Public policy writing that matters / David Chrisinger.
Description: Baltimore : Johns Hopkins University Press, 2017. | Includes bibliographical references and index.
Identifiers: LCCN 2016033367| ISBN 9781421422268 (paperback) | ISBN 9781421422275 (electronic) | ISBN 1421422263 (paperback)
Subjects: LCSH: Communication in public administration. | Communication in politics. | Written communication. | Persuasion (Rhetoric)—Political aspects. | BISAC: TECHNOLOGY & ENGINEERING / Technical Writing. | BUSINESS & ECONOMICS / Business Writing. | POLITICAL SCIENCE / Public Policy / General. | POLITICAL SCIENCE / Government / National. | REFERENCE / Writing Skills.
Classification: LCC JF1525.C59 C35 2017 | DDC 808.06/632—dc23
LC record available at https://lccn.loc.gov/2016033367

A catalog record for this book is available from the British Library.

Johns Hopkins University Press uses environmentally friendly book materials, including recycled text paper that is composed of at least 30 percent post-consumer waste, whenever possible.

PUBLIC POLICY WRITING THAT MATTERS

The scariest moment is always
just before you start. After that,
things can only get better.

—Stephen King

CONTENTS

ACKNOWLEDGMENTS

When I started working at the Government Accountability Office (GAO) in 2010, I never thought I would someday sit down to compose a thank-you to all those who have helped me write this book. I was an academic-in-training who was planning on waiting out the recession by working for the federal government. Two years. That's how long they said I had to commit, and that's how long I thought I would stay. After two years, I would return to graduate school, finish my PhD in German history, and land some amazing tenure-track job—or so I had planned. But then I fell in love with public policy, and I have Patrick DiBattista, my first boss and one of my most cherished mentors, to thank for that. From day one, Patrick gave me the confidence to advocate for better writing and storytelling. He believed in me and my abilities and was always there when I needed him. I probably wouldn't have made public policy writing my profession were it not for him.

I'd also like to thank Amy Buck, an analyst at GAO who has become a good friend and who helped me land the job that led to this book. After working with me on a report, Amy introduced me to Carey Borkoski, who was directing the Master of Public Policy Program at Johns Hopkins University, Amy's alma mater. Based on Amy's recommendation, Carey hired me to teach an intensive public policy writing course to the new graduate students. This book comes from the class I developed there. Thank you, Carey, for all the years of support.

There are others at GAO I cannot forget to mention. Barbara Bovbjerg, our fearless leader, has always been supportive and has pushed our team to be better storytellers. She's given me free rein to teach writing in a variety of nonconventional ways, including a series of "Schoolhouse Rock"–style training courses for government auditors. Thank you to Sue Bernstein and James Bennett and a whole host of others for helping make those courses a reality. I'd also like to thank Andy Sherrill and Kate van Gelder for trusting me and affording me this opportunity to do what I love. My fellow communications analysts—Holly Dye, Susan Aschoff, and Charlie Wilson—have also been encouraging and have taught me plenty about what it takes to write effective public policy. My sincerest thanks also go out to Dorothy Goldsmith, who runs the writing training at GAO, for believing in me and sending me all over the country to teach. Perhaps most importantly, though, I'd like to thank Dorothy for continually helping me work through different writing challenges and for helping me find new ways to improve the writing at GAO. Collin Fallon and Allison O'Neill have been great as co-instructors. They never mind when I pick their brains, and I appreciate their collaboration.

Almost three years ago, I had a kernel of an idea that eventually became this book. Before my plan was fully developed, I reached out to Johns Hopkins University Press. Fifteen minutes after I sent that fateful email, Kelley Squazzo answered my query and helped me put together a proposal. Before she left the Press, she championed my work, and I cannot thank her enough for all she did to make this book a reality. I'd also like to thank her successor, Robin Coleman, for his tireless support of this project. I've learned a great deal from him. In addition, this book has been greatly improved by the keen insights provided by my copyeditor, Julia Ridley Smith.

Lastly, I'd like to thank my beautiful and loving wife, Ashley, for so selflessly putting up with my absences and for always believing in me. I could not do this without her.

PUBLIC POLICY WRITING THAT MATTERS

INTRODUCTION

If you're reading this book, I'm willing to bet you're passionate about public policy and its power to make real, principled, and lasting change in the world. You also probably suspect that the way in which public policy is written matters as much—if not more—than the substance of the policy itself. As public policy professionals, we can have amazing ideas on how to improve the world, but if we aren't able to communicate these ideas well, they won't become reality. If you're new to this world of analysis, decision making, and communication for the public good, you might assume that policies gain support and are eventually implemented because of their methodological soundness or logic. While these things matter, the most effective professionals in this field are the ones who have mastered the power of persuasion. They are able to capture their readers' attention, and they are able to move their readers, to make them care enough to do something about the problems they've found. It is important to realize that the most difficult work public policy professionals do is not developing a point of view but rather getting someone in power to agree with that point of view. This book is intended to help you with both.

Even if you've been around the world of public policy for some time, you'll benefit from reading this book. If there's one thing I've learned about public policy, it's that writing almost always takes a back seat to sound analysis. We spend a huge amount of time trying to figure out *what* we want to say, which sometimes results in not having enough time to figure out *how* we want to say it. It's my hope that you'll read this book, dog-ear the pages that are most helpful to you, and pull it off the shelf each time you sit down to write a report or memo.

The central argument of this book is relatively simple: Writers of public policy are most effective and persuasive when they tell clear and concise stories. Short, smart, and simple are our goals. Clear and concise stories help make complex information and analyses accessible; they make them more interesting and persuasive, as well. Unfortunately, most public policy writing consists not of clear and concise stories but of muddled arguments and overly detailed explanations, which are far less persuasive. And even when there is a story to be told, many writers bury that story under jargon, complex sentence constructions, and incoherent paragraphs. Take as an example a sentence a colleague of mine read in the rough draft of a report he was helping edit a couple of years ago:

> The potential for inconsistent penalty administration within a decentralized management structure is exacerbated by the complexity of the penalty process within the IRS.

You don't have to look hard to find government reports riddled with such sentences. The problem, however, is that sentences like the one above don't tell stories readers can easily understand. Think about it for a moment. What story is the sentence above trying to tell? At its core, this sentence claims that "potential" is "exacerbated" by "complexity." Do me a favor. Try to picture with your mind's eye what *potential exacerbated by complexity* looks like. What do you see? Nothing? That's a problem. Readers better understand stories they can picture. It's that simple. If you take nothing else away from this book, let it be that: Write stories that your readers can picture in their mind's eye.

You may ask, What's so bad about such a sentence? You might argue that what the IRS does *is* complex, and sometimes it takes complex language to describe complex actions. Or you might decide that maybe the writer didn't want to come down too hard on the agency. The problem is that when most readers are confronted with a sentence like this, they're not going to take the time to deconstruct it, to make sense of the obfuscation. Some may blame themselves at first. They'll think they must have missed something and re-read the sentence a few times, but they won't do that for long. Soon enough, after reading more such sentences, they'll begin to blame you. They'll doubt what you have to say, and they won't be persuaded by your argument. When we make our readers work too hard, they give up on us—and our ideas.

What if instead we wrote:

> The IRS likely administers penalties inconsistently because it has a decentralized management structure and a complex penalty process.

Close your eyes again. Try to picture in your mind's eye *the IRS administers penalties*. It's much easier to picture, right? We know who is doing what, and we can expect now that the writer is going to explain to us why what we've just read is less than ideal.

Many psychologists and those who study how the brain works have known for quite some time that humans are primed for telling and understanding clear narratives. It's how we've survived as long as we have. Let's say, for example, that one of our distant ancestors was out for a walk along a river just as the sun was beginning to set. He was minding his own business, taking in the pristine views of the countryside. But then, out of nowhere, a saber-tooth tiger attacked. And let's say that this ancestor of ours was fortunate enough to survive the attack relatively unscathed and was able to make it back to the cave. I bet that once he regained his strength, he told his friends and family all about what had happened. He probably came to the conclusion that you should be careful at sundown while walking along the river, and that you should never let the beauty of the land distract you from checking your surroundings for predators. It's this kind of storytelling that has helped ensure wisdom is passed down in the hopes that other people and future generations might avoid making the same mistakes over and over again.

Now imagine how quickly our species might have died out had our ancestor buried the point of his story by using unclear subjects and verbs that showed no action. What if, instead of telling a story in which he made it clear how to avoid being eaten, he said something like, "The potential for untimely expiration by means of predator attack is exacerbated by the inability to maintain ocular awareness during twilight while in close proximity to a natural flowing watercourse"?

There's one more thing psychologists and cognitive researchers have learned about the way people process new information that is relevant to our work. They've found that information that seems random and disconnected is "costly" for your audience to obtain, store, manipulate, and retrieve. By costly, I mean that if your audience has to grapple with information to make sense of it, their brains are forced to work quite hard, and brains don't like to work unnecessarily hard. If the connections between points aren't clear, your audience has to create the connections themselves. Not only could these connections be wrong, but this work is tiring, and most readers won't stick with you for long. By telling good stories, however, we can help our audience process and retain information much more easily. We can do this by making facts and figures "sticky," that is, easy to pick up and hold on to.

Nassim Taleb is a philosopher and risk analyst whose work focuses on problems of randomness, probability, and uncertainty. In his 2007 book, *The Black Swan: The Impact of the Highly Improbable,* Taleb illustrates perfectly how to make our information stickier. He offers a useful example in which he asks his readers to consider two separate stories. The first story is "The king died and the queen died." He then asks us to compare that story to "The king died, and then the queen died of grief." These stories, he writes, show the distinction between a mere succession of events and a plot. "But notice the hitch here," he continues. "Although we added information to the second statement, we effectively reduced the dimension of the total. The second sentence is, in a way, much lighter to carry and easier to remember; we now have one single piece of information in place of two. As we can remember it with less effort, we can also sell it to others, that is, market it better as a packaged idea. This, in a nutshell, is the definition and function of a *narrative*." Above all else, your audience will find messages that are framed as stories easier to understand, more memorable, and, perhaps most importantly, more convincing.

If all this is true, why don't we instinctively write better? Haven't we evolved to tell stories? Shouldn't that mean that we have evolved to write them as well?

There are at least two reasons. First, public policy professionals tend to write muddled reports if the thinking behind the writing is also muddled. "Clear thinking becomes clear writing," says William Zinsser, author of *On Writing Well.* "One can't exist without the other."

I don't know who wrote that sentence about the IRS above, but I'm willing to bet that he or she was an intelligent person who spent weeks or even months learning as much as possible about the IRS and its decentralized management structure. The writer probably talked to a dozen or more agency officials and pored over hundreds of pages of internal reports and relevant legislation. What that person did not do, however, was take the time to make sense of the information and truly understand it. The result was that he or she wrote an unclear sentence.

The first purpose, then, of public policy writing that matters is to synthesize the current state of knowledge regarding your topic and write reports that can be understood and trusted as accurate. Put another way, your readers want to better understand a problem they do not have time to research on their own—though they may feel strongly about one side or the other—and they want to know what you think will fix that problem.

The capacity to synthesize information and make sense of it has become ever more crucial as information continues to be published at dizzying rates. The Internet alone contains an unimaginably vast amount of digital information, and readers simply don't have the time to process even a fraction of a percent of everything that is available on any given topic. Writers who are not able to sew together information from disparate sources into a coherent story are unlikely to make arguments that matter.

The second reason we struggle to write effectively and persuasively has to do with our audience. Who reads public policy? Policy makers, obviously, but academics, journalists, and other concerned people do as well. What do most of these readers have in common? For one thing, while they may be educated, most people who read public policy are probably not experts on the matter at hand. In addition, as I just mentioned, these readers probably do not have time to conduct in-depth research on their own. Lastly, while they may be skeptical, they ultimately want to know what you—the expert—has to say on the subject. The problem comes when the expert writes for those who are not. I've found that most of these writers struggle to predict what their readers will think is unclear. "What we write," say Joseph M. Williams and Gregory Colomb, authors of *Style: Lessons in Clarity and Grace,* "always seems clearer to us than to our readers, because we read into what we want them to get out of it. And so instead of revising our writing to meet their needs, we call it done the moment it meets ours."

The second purpose, then, of public policy writing that matters is to accurately and concisely communicate complex information, unbiased analysis, and logical and practical recommendations that meet the needs of a busy and skeptical reader. Without such effective and persuasive public policy writing, our policy makers won't have the easily understandable information they need to efficiently solve important problems that affect our society and the environment.

Thinking and writing clearly are, however, no easy tasks. They are skills that public policy professionals have been struggling to master probably for as long as there has been public policy. In his famous 1946 essay, "Politics and the English Language," George Orwell argued that the British government's writing was "lifeless" and "imitative." "In our time," he wrote, "it is broadly true that political writing is bad writing." What was Orwell's plan to fix such writing? "If you simplify your English," he wrote, "you are freed from the worst follies of orthodoxy."

Since Orwell's critique, plenty of people in the US government have tried to improve on the turgid prose that most of us struggle to comprehend. In 1966, John O'Hayre, an employee of the Bureau of Land Management at the time, published a book—*Gobbledygook Has Gotta Go*—in which he argued that convoluted writing had made government documents impossible to read. A few years after that, President Richard Nixon began requiring the *Federal Register*—a daily publication that issues proposed and final administrative regulations of federal agencies—to be written in "layman's terms." In 1998, President Bill Clinton issued an executive order requiring all federal employees to use short sentences; the active voice; and "common, everyday words." Most recently, President Barack Obama signed into law the Plain Writing Act of 2010. Aimed at improving "the effectiveness and accountability of Federal agencies," the Plain Writing Act promotes "clear Government communication that the public can understand and use." Above all else, this concerted effort to give the public information it can actually understand and use should help build trust in the government, something that has been sorely lacking in recent years. After all, how can you trust someone if you do not understand what they are saying?

Passing a few executive orders and a law here and there is not, however, enough to get public policy professionals to write clearly and concisely. That's where this book comes in. Through the discussion, examples of effective and ineffective public policy writing, and exercises—as well as helpful hints and cautionary tales—you will learn not only the principles of clear and concise public policy writing but also how to apply those principles to your own work.

In particular, you will learn how to

- develop, organize, and synthesize information and analytic results, ensuring that different sections of your writing logically link;
- present the results of your analysis as persuasively as possible;
- write deductively, with strong topic sentences and precise wording;
- write focused, unified, and coherent paragraphs, as well as effective sentences that are free of common errors;
- choose words that communicate meaning precisely and directly; and
- develop effective graphics that help you tell your story.

Much of the writing advice included in this book is, admittedly, not new. Some of it may be new to you, of course, and if it is, I hope you find it helpful. If, however, you are well versed in the study of effective writing, you will probably recognize the influence that a number of truly fantastic writing manuals have had on my development as a writer, editor, and teacher. In hopes of complementing such seminal writing texts as William Strunk Jr. and E. B. White's *The Elements of Style,* William Zinsser's *On Writing Well,* and Joseph M. Williams's *Style,* I have translated and synthesized their sage advice into practical tips, tricks, and tools directly related to the stylistic requirements and limitations of public policy writing. Even if you think you know all there is to know, I guarantee you will read things in this book that you either haven't seen before or haven't thought about in the ways I present.

Let's begin.

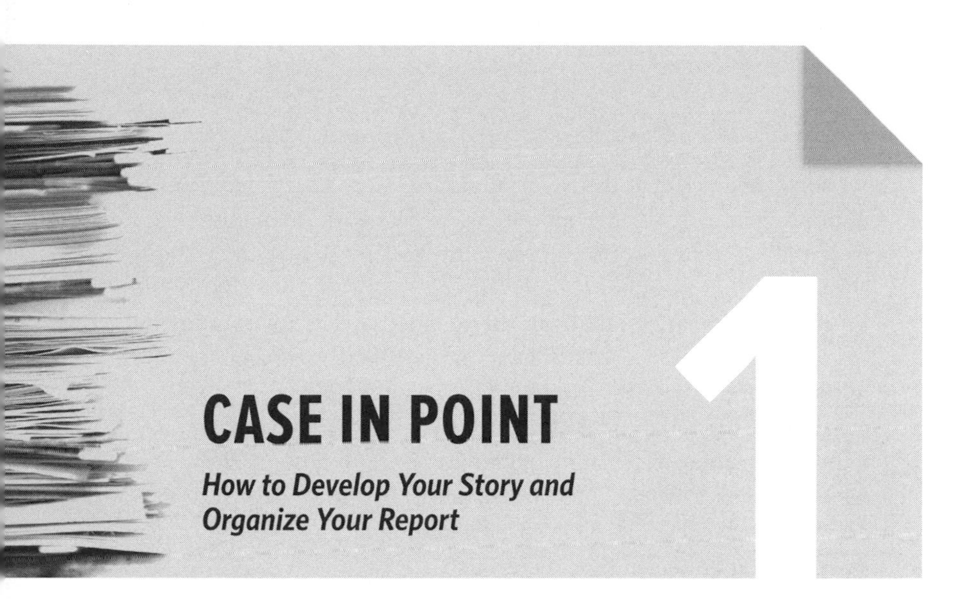

CASE IN POINT

How to Develop Your Story and Organize Your Report

In this chapter, you will learn how to develop your message, organize your findings, and prepare public policy writing that matters. As an example, we will closely examine the issue of suicide among US military veterans, as well as subsequent legislation that was passed into law in 2014, to explore ways public policy professionals and advocates have (1) analyzed research findings from a variety of sources, (2) organized them into a coherent structure, and (3) advocated for change using clear and concise stories. The example I deconstruct below has flaws, but it also has elements that are nearly perfect illustrations of the kinds of issues you will deal with as you write public policy. I will be clear about which are which.

Before we dive into the recommendations, I want to share some data and a little background information on suicide by military veterans. From there, I'm going to show you how the nonprofit advocacy group Iraq and Afghanistan Veterans of America moved from gathering data and background information to making policy recommendations to seeing its suggestions become law.

VETERAN SUICIDE

According to a January 2012 study released by the US Department of Veterans Affairs (VA), from 1999 through 2011, 22 military veterans killed themselves *every day*. That's about one suicide every 65 minutes for 11 years. After the study was released, however, the VA acknowledged that the reported figures were likely significantly underestimated, considering they were based on incomplete data from only 21 states, not including California or Texas—the 2 states with the highest number of veteran residents. (On July 7, 2016, the VA released updated data from the most comprehensive study on suicide among military veterans ever undertaken. After analyzing 50 million veterans' records from 1979 to 2014, including veterans from every state, the VA found that in 2014 about 7,400 veterans committed suicide, which works out to about 20 veterans per day. These new data were not, however, available when Iraq and Afghanistan Veterans of America began their advocacy for a national policy to help curb suicides among military veterans.)

Suicide among military veterans is not a new phenomenon. Statistics on suicide among Vietnam veterans, while sketchy, are still quite disturbing. Some have stated that around 100,000 Vietnam veterans have ended their own lives, but that statistic was extrapolated from a study that found the rate of suicide among institutionalized veterans was 23 percent higher than among other institutionalized patients in the same age group. In 1984, Myra MacPherson published perhaps one of the most definitive accounts of life after war for Vietnam veterans, *Long Time Passing,* and in it she quoted Dr. Jack Ewalt, the psychiatrist who headed the VA's mental health division. He said that despite the sketchy statistics, "We know there are a lot of suicides among veterans on the outside that never get reported to us." MacPherson also quotes Dr. Victor De Facio, a clinical psychologist who counseled Vietnam veterans. He says that suicide is a serious problem among Vietnam veterans and that "it's not only suicides you have to look at. Single-car accidents are often suicides. Or those covered up by the family—or drug overdoses. Alcoholism is another way to kill yourself. Another way is to simply rot. These are the veterans who have given up—the 'living dead.' The more time passes without giving assistance to veterans who need it, the more we lose."

Many researchers who have looked into the risk factors for suicide among military veterans have found that veterans have the same risk factors for suicide as the general population. Those factors include feelings of depression, hopelessness, post-traumatic stress disorder, a history of trauma, and access to firearms. Dr. Jan Kemp, for example, the VA national suicide prevention coordinator, has stated for years that "veterans who die by suicide look a lot like Americans who die by suicide."

Such anecdotal observations are now being validated by generalizable data. A 2013 long-range study of more than 150,000 active and retired military personnel was published in the *Journal of the American Medical Association.* It found no proof that the rising suicide rate among veterans *directly* stems from combat in Iraq and Afghanistan. Instead, the researchers say, the risk factors for suicide in the military are the same as those in the civilian world: (1) depression, (2) drinking problems, and (3) being a man. More specifically, the researchers found that depression more than doubled the risk of suicide, as did alcohol abuse. In addition, men were twice as likely as women to commit suicide. The researchers did not, however, find a clear link between suicide and the number of deployments to Afghanistan or Iraq, or to combat exposure. This "does not prove that deployment has nothing to do with suicide risk," says Dr. Nancy Crum-Cianflone of the Naval Health Research Center. It simply means "deployment was not related to the risk of suicide."

There are a staggering number of variables that can contribute to such broad social issues, and the effect of a policy may vary greatly among communities based on their specific resources or backgrounds. That doesn't mean, however, that we should give up trying to find a policy solution to veteran suicide. In the years ahead, hundreds of thousands of war-weary men and women—as well as those who will never deploy—are expected to leave the military and be confronted by the stresses of civilian life. They deserve better than to feel abandoned by a country many of them will have sacrificed so selflessly to defend.

Now that we have a better understanding of the issues at hand, let's take a look at what the Department of Veterans Affairs is doing to address veteran suicides.

VA's Suicide Prevention Efforts

The Department of Veterans Affairs' mission is to fulfill President Abraham Lincoln's promise to "care for him who shall have borne the battle, and for his widow, and his orphan" by serving and honoring America's veterans. Even though the VA was never designed to reintegrate veterans back into society or help them repair social relationships, the agency has made it its new mission to prevent injury, reduce disease, and increase the quality of life for eligible veterans, which requires accessible, efficient, and quality care.

To update the US Congress on what the VA does to help veterans with mental health conditions, particularly those who are at risk for suicide, Dr. Harold Kudler, the chief mental health consultant at the Veterans Health Administration (VHA) within the VA, testified before the Senate Committee on Veterans' Affairs on November 19, 2014. Among other things, Kudler said:

- *More veterans are receiving care through the VA:* "The number of Veterans receiving specialized mental health treatment from VA has risen each year, from 927,052 in fiscal year 2006 to more than 1.4 million in fiscal year 2013."
- *Better data are being collected:* "In 2010, DOD [the Department of Defense] and VA approved plans for a Joint Suicide Data Repository (SDR) as a shared resource for improving our understanding of patterns and characteristics of suicide among Veterans and Servicemembers. The combined DOD and VA search of data available in the National Death Index represents the single largest mortality search of a population with a history of military service on record. The DOD/VA Joint SDR is overseen by the Defense Suicide Prevention Office and VA's Suicide Prevention Program."
- *Those who need it are receiving help with transitioning their mental health care from the military to the VA:* "All Servicemembers leaving the military who are receiving care for mental health conditions are automatically enrolled in the *in Transition* program. In this program, trained mental health professionals assist these Servicemembers through 'warm handoffs' to new care teams in VA or in the community. Further, mental health medications prescribed by clinicians from [DOD] will be carried over into VA care unless a specific safety or clinical reason to make a change is identified."

- *More mental health employees have been hired:* "VA has added 2,444 mental health full-time equivalent employees and hired over 900 peer specialists and apprentices."
- *Veterans have many ways of accessing health care:* "VA has many entry points for VHA mental health care. These entry points include 150 medical centers, 820 Community Based Outpatient Clinics (CBOC), 300 Vet Centers providing re-adjustment counseling, a Veterans Crisis Line, VA staff on college and university campuses, and other outreach efforts."
- *Mental health services have been expanded:* "As of April 2014, VHA has 21,158 Mental Health full-time equivalent employees providing direct inpatient and outpatient mental health care. VA has expanded access to mental health services with longer clinic hours, telemental health capability to deliver services, and standards that mandate immediate access to mental health services to Veterans in crisis."

In addition, the VA's budget has nearly tripled since the terrorist attacks of September 11, 2001, the DOD's medical system budget has nearly quadrupled, the VA's disability claims process has been fast-tracked, and the federal government has launched a multi-billion-dollar effort to improve the VA's IT infrastructure.

Despite all of these policy changes, many do not believe the VA is doing enough to stem the tide of veteran suicides. "With more than 22 veterans a day still dying from suicide year after year," wrote Alex Nicholson, the Legislative Director for Iraq and Afghanistan Veterans of America (IAVA), "it is insufficient to say that the Department of Veteran Affairs already has the resources and focus to fight this ongoing crisis. If it did, we would have already seen a marked decrease in veteran suicides instead of a steady continuation of the epidemic."

After surveying more than 2,000 of its veteran members in 2014, IAVA reported that:

- 31 percent indicated they have thought about committing suicide since joining the military,
- 40 percent know at least one post-9/11 veteran who has *committed* suicide, and

- 47 percent know at least one post-9/11 veteran who has *attempted* suicide.

In response to these results, IAVA decided to tell legislators a story about a war hero who fell through the cracks before he could be saved.

The Clay Hunt Suicide Prevention for American Veterans (SAV) Act of 2014

Clay W. Hunt was a decorated combat veteran of Iraq and Afghanistan who went on to become a prominent advocate for veterans suffering from post-traumatic stress. Despite his efforts to find new meaning and purpose as a civilian, on March 31, 2011, he locked himself in his apartment and ended his debilitating battle with post-traumatic stress, depression, and survivor's guilt with a self-inflicted gunshot. Hunt's death helped galvanize IAVA's effort to provide better help for suicidal veterans. "The message I've been trying to convey to people," says Paul Rieckhoff, the founder and CEO of IAVA, "is that if this can happen to Clay Hunt, it can happen to anyone. He was involved. He had a supportive family. He was going to the VA. He was doing the right things. And it still happened."

"Clay enlisted in the Marine Corps in May 2005 and served in the infantry," Clay's mother, Susan Selke, told the Senate Committee on Veterans' Affairs on November 19, 2014. "In January of 2007, Clay deployed to Iraq's Anbar Province, close to Fallujah. Shortly after arriving in Iraq, Clay was shot through the wrist by a sniper's bullet that barely missed his head. After he returned [home] to recuperate, Clay began experiencing many symptoms of post-traumatic stress, including panic attacks, and was diagnosed with PTS [post-traumatic stress] later that year."

In 2008, after completing Scout Sniper training, Clay deployed again, this time to southern Afghanistan, where he lost two friends. After he left the Marine Corps in 2009, Clay sought help for his problems with depression and stress. He saw multiple doctors and got medication for his panic attacks and depression, though he struggled to get disability payments after the VA misplaced his paperwork. Susan said:

He received counseling only as far as a brief discussion regarding whether the medication he was prescribed was working or not. If it was not, he would be given a new medication. Clay used to say, "I'm a guinea pig for drugs. They'll put me on one thing, I'll have side effects, and then they put me on something else."

In late 2010, Clay moved briefly to Grand Junction, Colorado, where he also used the VA there, and then finally home to Houston to be closer to our family. The Houston VA would not refill prescriptions Clay had received from the Grand Junction VA because they said that prescriptions were not transferable and a new assessment would have to be done before his medications could be re-prescribed.

Clay only had two appointments in January and February of 2011, and neither was with a psychiatrist. It wasn't until March 15th that Clay was finally able to see a psychiatrist at the Houston VA medical center. But after the appointment, Clay called me on his way home and said, "Mom, I can't go back there. The VA is way too stressful and not a place I can go. I'll have to find a vet center or something."

Just two weeks after his appointment with a psychiatrist at the Houston VA medical center, Clay took his own life.

"Not one more veteran should have to go through what Clay went through with the VA after returning home from war," Susan concluded. "Not one more parent should have to testify before a congressional committee to compel the VA to fulfill its responsibilities to those who served and sacrificed."

The story of Clay Hunt's life and untimely death had all the elements needed to make it a relatively easy policy story to tell. Hunt was a decorated combat veteran who fought for his country and was wounded as a result. He became a Scout Sniper, an elite specialty in the US Marine Corps. In addition, Hunt had the strength and wisdom to seek help when he wasn't doing well. He went to the VA, but the VA failed him: so the story goes. At the very least, when he tried to get help, the VA wasn't able to help him.

The question then becomes: What policy changes could be made to ensure that other veterans don't meet the same fate as Clay Hunt?

After IAVA spearheaded its creation, the Clay Hunt SAV Bill was introduced in the US Senate on November 17, 2014. "Combating veteran suicide has been a top priority for IAVA this year," said Rieckhoff. "When passed, this bill will ensure our veterans receive the top quality mental health care they deserve."

Specifically, the Clay Hunt SAV Bill would have

- required independent evaluations of all mental health care and suicide prevention programs provided through the departments of Defense and Veterans Affairs,
- authorized the VA to collaborate with Veteran Service Organizations and nonprofit mental health organizations to prevent suicide among veterans,
- established a drug take-back program for prescription drugs at VA medical facilities, and
- provided up to $120,000 per year in student loan repayments to recruit psychiatrists who commit to working for the VA.

Instead of passing the $22 million bill, however, Republican senator Tom Coburn blocked the bill from being considered by the Senate. "I don't think this bill would do the first thing to change what's happening" in terms of veteran suicide, he said. "I'm going to be objecting to this bill because it actually throws money away." Rajiv Jain, the Veterans Health Administration's assistant deputy under secretary for health for patient care services, agreed with Coburn. He told members of the House Veterans' Affairs Committee that while the VA supported the goals of the bill, the legislation overlapped with existing programs.

Others criticized the bill for not going far enough. Some argued, for example, that the bill didn't address the concern that the VA wasn't doing enough to include spouses and families in treatment, nor did it address the amount of medications, especially antidepressants, that are prescribed to military veterans. Others said that alcohol and access to firearms needed to be addressed. Mark Kaplan, a professor of social welfare at the University of California Los Angeles' Luskin School of Public Affairs, said that he saw a "perfect storm" when alcohol and guns are both present. Kaplan, who studies suicide risk, said, "I think some of those acts are impulsive. I think they are faced with a precipitating crisis, they drink excessively and may not have a history of chronic alcohol problems, and they simply drink to be able to pull that trigger."

In the end, Senator Coburn's stance turned out to be purely ceremonial. After clearing the House of Representatives unanimously, the Senate approved the Clay Hunt SAV Bill on February 3, 2015, by a 99-to-0 vote. Senator Coburn retired a couple of weeks before the votes were cast. The bill, which was subsequently signed into law by the president, changed slightly and now calls for

- increased access to mental health care by, among other things, creating a peer support and community outreach pilot program to assist transitioning service members as well as a one-stop, interactive website of available resources;
- a pilot program to repay the loan debt of students in psychiatry so it is easier to recruit them to work at the VA; and
- an annual and independent evaluation of VA mental health and suicide-prevention programs that will boost VA's accountability of mental health care.

In response, the Department of Veterans Affairs convened a summit in February 2016 at which it vowed to make a number of changes. Specifically, the officials present stated in a press release that the department planned to, among other things, (1) evaluate its suicide prevention programs, (2) set a goal for same-day evaluations of those needing mental health care, (3) commission a study looking at the impact of deployments on risk of suicide, (4) use data to conduct predictive modeling, and (5) expand telemental health care for veterans who don't live near a VA medical center. Clearly, the story IAVA told about Clay Hunt mattered. It mattered so much so that it inspired nearly unanimous support for change from a bitterly partisan governing body.

So how was IAVA able to match deficiencies in the ways the VA was serving veterans with persuasive and practical policy recommendations? Let's take a closer look.

DEVELOPING YOUR MESSAGE

The stories you tell must appeal to your readers. You must continually remind yourself that your readers are human and will focus on ideas and evidence that support their values, beliefs, and preconceived worldviews. Your readers will also likely ignore anything that contradicts those things. Think about the 24-hour news cycle as an example—it's story time for adults. Facts and evidence are provided, but they have generally been handpicked to match a partisan perspective. Half the story is often left out, and we don't always notice because the stories we hear make sense to us.

Why did the story IAVA told not resonate with Senator Tom Coburn? Like I said above, Clay Hunt was a highly trained, decorated, wounded warrior who sacrificed much for his country. And when he sought help dealing with his traumatic experiences, the VA wasn't able to help. Why wasn't this story good enough to persuade Coburn to take action? The answer is pretty simple. Senator Coburn was well known for being a fiscal and social conservative who created a name for himself by opposing deficit spending and pork-barrel projects. His Democratic colleagues, in fact, nicknamed him "Dr. No." It's no wonder he wasn't persuaded by Hunt's story. Yes, a decorated combat veteran killed himself, and yes, that is tragic. But in Coburn's eyes, simply spending more money wasn't going to prevent something like that from happening again. Because IAVA didn't do enough to appeal to Senator Coburn's values, their evidence didn't matter much to him.

But what about the other 99 senators? Hunt's story speaks to equal opportunity, achievement and success, practicality and efficiency, progress, and freedom. A young man volunteered to sacrifice himself in service to his country and to protect American freedoms, performed admirably on the field of battle, was wounded, struggled to make sense of life after war, and was failed by the people who pledged to honor and serve him. Such a story flies in the face of what those 99 senators and 435 House representatives probably valued most. That something must be done was the only logical conclusion for people who share the worldview that citizens like Hunt deserve to be cared for and supported.

Let's take a closer look at why people make decisions and at what we know about making a sound argument. As I mentioned above, the details of Hunt's story are important in that IAVA used each detail to appeal to the values and beliefs of the majority of people who were going to be on the receiving end of the story. For decades, sociologists have studied what Americans value. According to sociologist John J. Macionis, the author of *Sociology,* a seminal textbook, the core American values are:

- equal opportunity,
- achievement and success,
- material comfort,
- activity and work,
- practicality and efficiency,
- progress,
- science,
- democracy and enterprise, and
- freedom.

Appealing to your audience's values and worldviews to persuade them to take action is nothing new, by the way. Over 2,000 years ago, Aristotle laid out the three pillars of rhetoric that must be employed to persuade others: ethos, pathos, and logos.

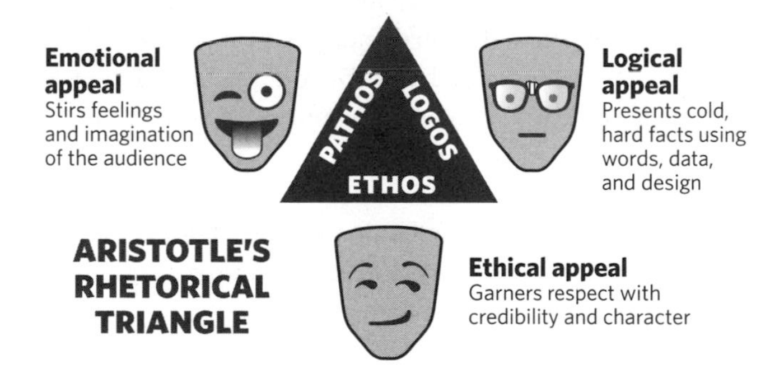

IAVA has a reputation for being sincere, truthful, and rational in their dealings with policy makers (ethos), and they were able to use Hunt's story to stir the emotions of their audience (pathos). They also presented evidence and data to support the conclusions they had reached (logos). Aristotle would have been quite proud, I'm sure.

This all seems so manipulative, you might think. In a sense, it certainly is. You, the public policy professional, must convince a busy and skeptical audience to agree with what you've found and recommend. Rigorous analysis and evaluation alone will not get that job done. You'll also need to establish credibility and stir your reader's emotions, and the easiest way to do both of those things is to communicate stories as clearly and concisely as possible.

While telling Hunt's story was a necessary first step—and undoubtedly helped pave the way for change—it wasn't enough on its own to inspire Congress to act. IAVA had to use Hunt's story to start a conversation and serve as a hook to lead the legislators toward accepting their policy recommendations.

Let's take a look at how IAVA might have come up with their recommendations.

The Four Elements of a Finding

In this section, I introduce and explain the four essential parts of a public policy argument: condition, criteria, cause, and effect.

The questions IAVA decided to answer determined how they organized their work and presented their recommendations. Such is true for all public policy writing that matters. For the purposes of this section, let's assume that you are the legislative director for IAVA. The questions you decide to answer will (1) define what your report must address, (2) help you decide what to include and what to leave out, and (3) provide a road map for your reader to follow.

Let's also assume that you began your research with the following question:

> How prevalent are suicides among American military veterans, and what can be done to help prevent them from occurring in the future?

One way that many public policy professionals begin to organize their findings (answers to questions) is by putting what they have found into four distinct "buckets" or elements. According to "generally accepted government auditing standards (GAGAS)," which were developed by the Government Accountability Office and published by the Comptroller General of the United States, "auditors [that's you] should plan and perform procedures to develop the elements of the findings [answers to your questions] that are relevant and necessary to achieve the audit objectives."

Each of the elements serves a specific purpose and should not be left out. If you leave an element out of your report, your reader will be left hanging.

The *Four Elements of a Finding* are:

1. *Condition:* Answers the question, "What is happening?"

The condition identifies "what is" and describes the circumstances that have been observed and documented during the engagement. All reports should include a description of what is happening. Sometimes the condition will be good, and other times the condition is not ideal. In such cases, we call that finding a *deficiency.* If you find a deficiency, your reader will want to know what you propose to fix it. If there is no deficiency, you won't need to make a recommendation.

For example: At least 22 military veterans commit suicide every day. That's what's happening. At the same time, the VA offers a wide range of medical and health services, educational programs, and transition assistance for the nearly 22 million military veterans who live in the United States, 1.4 million military service members, and their families. This second condition is much more descriptive, thus helping you to set the scene. In terms of evaluating its own performance, the VA employs a number of methods, including program performance reviews, veteran satisfaction surveys, and periodic longitudinal reviews of certain veteran groups. However, the VA's suicide prevention and other mental health programs are not regularly and independently evaluated to determine whether the most effective approaches are being used to achieve program goals. Our main condition—our deficiency—is that last condition. Our recommendation will address the need for regular, independent evaluations of the VA's suicide prevention efforts and other mental health programs.

In my experience, public policy professionals spend the bulk of their time figuring out what is happening—up to 75 percent of their time, by some estimations.

2. *Criteria:* Answers the question, "What should be happening?"

It's not enough to point out a problem and say it's a problem. To determine what *should* be happening, you need criteria, which provide a context for understanding your findings. Criteria may be found in "laws, regulations, contracts, grant agreements, standards, measures, expected performance, defined business practices, and benchmarks against which performance is compared or evaluated." Essentially, criteria help you advocate for change using stronger logic. Your criteria should be reasonable, attainable, and relevant to the matters being investigated.

In addition, the criteria you use should be valid for the intended purpose. You need to be aware of how circumstances may have changed since the criteria were established.

For our purposes, our criteria are: The VA's mission is to help returning veterans get the help they need to deal with difficulties resulting from their service in the military. At the same time, because the VA's accountability has been called into question, regular independent reviews of its suicide prevention efforts and other mental health programs need to be conducted. After all, veterans should receive help from a VA that can be trusted to be high quality, accessible, and efficient.

3. *Cause:* Answers the question, "Why is the condition happening?"

Cause is the reason something happened or did not happen. In other words, cause is what produced the condition. Identifying cause may mean identifying the underlying reason why things are not working as expected, or cause may include the reason an intervention or program—or physical, social, or economic conditions—caused something to happen or not happen and the extent to which the intervention, program, or condition is the cause of the changes.

Sometimes what we initially think is a cause is actually a *result* of the condition. For example, one reason why veterans commit suicide may be that they don't have access to programs that have proven results. That makes sense. But what prevents the VA from providing access to programs with proven results? The answer to that question is the true cause. One cause may be that some programs offered by the VA have not been rigorously evaluated because the data needed to conduct such evaluations are difficult and time consuming to collect and report. As a result of that cause, the agency may choose not to evaluate its programs regularly and may not know whether the programs can be effective. When you find a cause that is actually the result of the condition, you'll need to dig a little deeper. Why, for example, doesn't the VA prioritize some of its funding for program evaluation? Is the VA incentivized to "take action" quickly over thoughtfully planning out and evaluating its efforts? Are the political consequences worse for not doing something or for not evaluating what you did? Good public policy professionals will not be satisfied with the first cause they find. They'll dig deeper and figure out whether there's something else going on.

Let's look at cause another way. If I walked out to the road in front of my house and saw a horse standing there, my first thought would be, "What is this horse doing in front of my house?" The answer to this question is the cause, but, like I said above, not all causes are created equal—or are even actual causes. After thinking for a minute or two about this horse situation, we could conclude that the reason the horse is in the road is because there isn't anyone guarding the road to ensure that a horse won't wander into it. There's a logic to that conclusion, but it's not necessarily the strongest logic. Or we could conclude that the reason the horse is in the road is because someone left the gate to his pen open, and he wandered into the road. We're probably getting closer to the root cause with this line of thinking. And if we drill down a bit further, we might conclude that the reason—the cause—the horse is in the road is because the farmer down the road from my house forgot to latch the gate and didn't check the pen regularly. As we can see, sometimes the cause is an issue related to prevention. Other times, the party in question failed to detect a problem, and in other cases, the issue lies with a failure to correct the problem. Your job as a public policy professional is to determine what the true cause is so that you can recommend a solution that will negate that cause and stop whatever negative thing is happening from continuing into the future.

The problem, however, is that when many public policy professionals detect a problem, they have so little time to figure out what caused it that they mostly rely on the agency or entity they are researching to determine what the true cause is. Let's imagine that you were tasked with determining the reason why there was a horse in my front yard. After a little analysis, you'd figure out that the horse came from the farm down the road. Now, if the farmer is anything like agency officials I've encountered in my work as a public policy professional, he likely won't admit to having left the gate open. "It was the weather," he might say. "The wind blew the gate open, and I cannot control the weather, you see, so this is not my fault." Or maybe he'd claim that he didn't have enough money to make sure his gate was adequate and if only the government would give him more money, then he could do his job effectively. In the course of your work, you'll need to dig a little deeper and probe a little harder to figure out what's really going on.

Think of it this way: Can you imagine how unsatisfied you'd be if you were sick and you went to the doctor to figure out the cause of your sickness, and all your doctor did was ask you why you thought you were sick? "Well, I ate a gas-station burrito yesterday for lunch," you might say. The doctor would then answer, "That must be it! The solution? No more burritos for you!" The bad burrito might be the cause of your sickness, but what if it's not? This example may seem silly, but I've seen plenty of public policy professionals fall victim to such thinking. As professionals, we need to be running the tests and asking the hard questions, all in an attempt to dig down to the root cause. Only then can we feel reasonably assured that our recommendation will actually help.

4. *Effect:* Answers the question, "What will happen if the status quo is maintained?"

If you want readers to take your recommendations seriously, you'll have to make sure the effect you present will convincingly demonstrate that change is needed.

For example, if the VA's suicide prevention efforts are not regularly and independently evaluated, the public may lose faith in the VA's efforts, and as long as veterans are treated by potentially ineffective programs, they may continue committing suicide at a rate far higher than their civilian counterparts.

Remember too that effects don't always have to be negative. Sometimes it may be best to point to the positive effects of making a change. Use your best judgment and decide whether positive or negative effects are best for your purposes.

* * *

Here are two rules about using the Four Elements of a Finding.

1. *Don't overwhelm your reader with too many conditions:* While it may be necessary to provide several conditions to set the stage for the rest of your recommendation, try not to present too many deficiencies all at once. If you include too many deficiencies, you'll have to include multiple sets of criteria, causes, and effects, and you may overwhelm your reader. Consider finding what you believe to be the most important condition(s) and develop them fully. After all, a solidly convincing policy recommendation beats three underdeveloped recommendations every time.

2. *The elements must match:* Your recommendation must match the condition and cause. Put simply, your recommendation must address the cause and resolve the condition.

Take a look at table 1.1 to see how you, the legislative director, might develop a recommendation that will hopefully stem the tide of veteran suicides.

Table 1.1. THE FOUR ELEMENTS OF A FINDING

Condition: *What is happening?*

- At least 22 military veterans commit suicide every day.
- The VA offers a wide range of medical and health services, educational programs, and transition assistance for 22 million military veterans, 1.4 million military service members, and their families.
- The VA employs a number of methods to help it determine how well its programs serve veterans, including program performance reviews, veteran satisfaction surveys, and periodic longitudinal reviews of certain veteran groups.
- The VA's suicide prevention and other mental health programs are not regularly and independently evaluated to determine whether the most effective approaches are being used to achieve program goals.

(*continued*)

Table 1.1. *(continued)*

Criteria: *What should be happening?*

- The VA's mission is to help returning veterans get the help they need to deal with difficulties they may have incurred as a result of their service in the military.
- Veterans should receive help from a VA that can be trusted to be high quality, accessible, and efficient.
- Because the VA's accountability has been called into question, regular, independent reviews of its suicide prevention efforts and other mental health programs need to be conducted.

Cause: *Why is it happening?*

- It is difficult, time consuming, and costly to collect and report qualitative and quantitative performance measurements needed to conduct full performance evaluations.
- It may not be a priority for the VA to collect such data and conduct such reviews.
- The VA is not currently required to have regular, independent evaluations conducted on its suicide prevention efforts and mental health programs.

Effect: *What is the positive or negative effect?*

- If the VA's suicide prevention efforts are not regularly and independently evaluated, the public may lose faith in the VA's efforts.
- As long as veterans are treated by potentially ineffective programs, they may continue committing suicide at a rate far higher than their civilian counterparts.

What the Journalist Knows

As we've seen above, public policy writers can focus on a narrow set of variables. They compare what's *actually* happening (condition) to what the rules say *should* be happening (criteria). Only then do they begin to search for reasons that things aren't working according to plan (cause). A good audit will also suggest ways to fix any shortcomings (recommendation).

In addition to the four elements of a finding, there is at least one other way you could approach your work so that you can be certain you're telling the whole story. Just like you, reporters don't have time to pontificate. New journalists are taught an effective tool for quickly sifting through information: the "Five Ws and an H" questions. Similar to determining the Four Elements of a Finding, answering the "who, what, when, where, why, and how" of an event forces you to include all the information readers need to put that event in context.

The last two questions—why and how—are especially important because answering them moves the report into the realm of analysis. After all, you have to know how and why something happens before there is any hope of changing it. Filtering the gathered information through the six questions is, in turn, a good place to start coming to terms with what can be written about a policy or program.

The "Five Ws and an H" Questions:

Who? The players in a program or process can be individuals or larger groups (such as departments or entire agencies).

What? The measurements and data that result from this question (dollars spent, numbers served, tons harvested, etc.) are sources of the facts and data that will allow for analysis later. The answers to this question also include telling anecdotal information developed from direct observation (what it's like to be there) and interviews (what people said about it).

When? The period of time to be included. Too short a time period risks removing context, but too long a time period can quickly dilute a revealing trend.

Where? Geography is not always an important factor, but it's worth considering whether location has an effect on a process or trend.

Why? Answering this question requires getting past any finger-pointing to discover the cause of a problem.

How? A sufficient understanding of any processes under review is impossible if you are not able to suggest improvements on how to make something better.

Checking the assembled information against the six questions is a great way to guard against missing data and gaps in understanding. It can also suggest areas in need of additional research:

> Why are most veterans who commit suicide close to retirement age, decades after their service ended?

It's possible that the reason behind a particular anomaly won't have a significant effect on the overall story, but readers will wonder what else you missed if obvious patterns aren't explained.

Indeed, following up on unexpected anomalies is a critical part of a thorough analysis. It is, however, only a first step. True insight begins when the entire dataset is probed with broader "how" and "why" questions.

> Are new programs and policies being targeted to veterans who are more likely to commit suicide?
>
> Why is spending going up while the number of veteran suicides remains fairly constant?

The important questions will be different on every project, but systematically "interviewing" your data with thoughtful questions is the surest way to coax a compelling story from a jumble of facts and figures.

WRITING A PUBLIC POLICY REPORT

Having a story to tell and knowing the elements of your finding are only half the battle. How the story is structured and presented also has a profound effect on the reader's experience. Above all else, it's best to keep the reader's needs in mind.

Once you know what story you want to tell, the next step is to organize your report (and your argument) in a way that seems logical to the reader.

The "Anatomy" and "Physiology" of a Public Policy Report

Explaining how an effective public policy report should be structured, a colleague of mine said that it was helpful to think in terms of anatomy and physiology. I think that's a brilliant way to make sense of public policy writing that matters. Anatomy, of course, is the study of living things and their parts. Physiology, on the other hand, is the more complicated study of how those individual parts function complementary to one another. When it comes to public policy writing, you need to know the anatomy (the parts) of a persuasive report, as well as the physiology (how the parts work together).

Here is a list of the main parts (the anatomy) of an effective public policy report.

Title: Reflects the bottom-line message. For example: "Veteran Suicide: VA Needs to Have Its Programs Independently Evaluated."

Executive Summary: Provides a short summary of your report, including your recommendation(s).

Opening Paragraphs: Set the stage. The opening paragraphs—no more than two or three, depending on the complexity of the issue—prepare the reader for the rest of the report. These first few paragraphs include (1) an explanation—a hook—for why this work was done, and (2) enough background information to ensure a reader will understand the answers to your research questions. Many call the opening "the hook," though I don't think that term accurately describes the real job of an introduction. Some journalists call the hook a "teaser":

> Scientists have discovered something that will change the way we think about everything.

Such a beginning is supposed to draw the reader in, usually with a personal vignette or some eyebrow-raising fact. But if we merely cite the astonishing dollar amount spent on a program, for example, we haven't done the job of preparing the reader for the rest of the report. And if we tell the reader all about how a program is run or about its legislative origins, we still haven't done anything that matters. We've just wasted the reader's time and attention.

Readers will become more knowledgeable about the issue and the program as you go further into the report. If you flood them with information before you say why you're giving it to them, they will have no reason to hang onto it.

Effective introductions in public policy writing begin with an *instability*—something new has happened or something has changed. Put simply: There is a problem.

Have there been changes in laws or in regulations, the effects of which are unknown? Has there been an event or controversy about which more information is needed if there is to be any resolution? What don't we know? Why might we need to know it? Why should the reader care?

An effective introduction connects the *instability* to some kind of *consequence* that matters to the reader. Simple as that. Take another look at my introduction to veteran suicide to see what I'm talking about.

Ultimately, you should be able to explain the premise for your research in one paragraph. Additional paragraphs might be needed for a more complex issue, but generally if you can't state the reason for your research within the first few sentences, you haven't got a handle on it. Don't tell the reader that something has raised concerns without at least hinting what those concerns are. The reader wants to know right away why he or she should care.

Background: Gives the context. For the reader who wants additional information, the background includes the contextual information needed to understand the report.

The background serves two functions: First, it offers the reader additional context—just enough information about the subject at hand so the reader can generally recognize the players, their roles, and their relationships. It's really just an elaboration of the introduction. It might include prior findings, trends, or more about the known current conditions. Second, the background establishes the criteria for making your evaluation. That's all the background does.

You can shorten almost any report by keeping the background to a few pages and using a writing style that resembles a talking voice. The background should not require the kind of technical, legalistic details or jargon that might be required for the findings in the main body of the report or, more likely, in the footnotes or appendix where your real documentation occurs. Nor is the background a manual on how to run the program or—even worse—a reference library of miscellaneous information.

In terms of sequence, the background should start with the program or the situation that prompted your investigation, not with a chronology that begins with something overly broad, like "Ever since the days of classical Athens . . ."

Headings: Identify key information. Headings are message driven; they state a finding in either a descriptive or evaluative way.

Your readers must know where one section stops and the next begins. You can use headings to identify the start of a new section. The most effective headings completely answer the question you pose in that section.

Some inexperienced writers think that if they reveal their main point early in the report, readers will be bored and not continue reading. That is not true. If you ask an interesting question, readers will want to see how you support your answer.

Headings should help readers find information quickly, leaving the details for the body of the report. To write concise but informative headings, use the following techniques:

Be informative, not wordy. Headings should convey the main point quickly. They do not need to repeat every word of the objective or touch on every point in the section.

Use second-level headings to elaborate on a first-level heading. Second-level headings should work together to outline the logical organization of the section. They should elaborate on, rather than repeat, information that is already clear from the first-level heading.

Save the details for the topic sentence. The focus of the heading should be on the main point, the answer to your question. Details that explain the answer should be moved into the topic sentence.

Use topical headings when appropriate. Topical headings are those that only describe the topic of a report section and do not contain a subject and verb. They are appropriate when a sentence-style heading would be long and cumbersome. This may be the case when the findings in an objective are so broad or disparate that there is no unified message.

Findings: Provide the evidence and analysis. Each section answers a specific, corresponding question. Such linkage (1) helps ensure that the reader will be able to follow any question throughout the report and (2) enables different readers with different informational needs to navigate your report. In other words, your readers will be able to easily select the issues they want to read about and skip over the information they don't want to read.

There is often the greatest confusion here about what to include and in what order. Ideally, each finding section

- presents a bottom-line answer to each question,
- breaks the finding down into its component parts, and
- offers just enough supporting evidence to prove your point.

I am sorry to say that I can't give you any simpler rule on deciding what's relevant and what isn't. I can tell you, however, that sentences are relevant to a point when they offer

- background or context;
- reasons supporting a point;
- evidence, facts, or data supporting a reason;
- an explanation of reasoning or methods; and
- consideration of other points of view.

If you need to refer to criteria in the finding section, do it gracefully. It's really a question of balance. There are ways to construct your sentences so that you emphasize what you found relative to what would be reasonable or desirable (the criteria).

The trick to focusing on the key purpose of any section, particularly the finding section, is to devote the main clause of most sentences to the subject at hand. You can weave ancillary information into subordinate clauses and phrases. Think of the main clause as the road. You want to keep yourself and your readers on that road, knowing where they are headed. Take this sentence, for example:

Draft: The April 2012 Executive Order required the VA and other agencies to prevent abusive and deceptive recruiting practices that target the recipients of federal military and veterans education benefits by strengthening their enforcement and compliance efforts.

In this revision, the main clause focuses on the finding:

Revision: In 2014, the VA launched a new system that allows veterans to file complaints of abusive and misleading school recruiting practices and other violations of the Principles of Excellence, as required by an April 2012 Executive Order.

Another thing I see public policy writers do frequently is lead with their research methods. The research you conducted is not the story, so don't lead with it. For example:

Draft: We analyzed *a* and *b*, and based on these thresholds it would be expected that *x* and *y* would happen, but this was not the case. Two factors that can explain this gap are exemptions from filing reports in each program, as well as failure to file.

Instead, lead with the revelations (the story):

Revision: Exemptions and failure to file required reports also obscure the actual number of facilities, as indicated by our examination of reports filed for two different programs. [Follow this sentence with a discussion of that examination.]

Above all else, don't get ahead of your story. Don't rush to judgment. Don't write about the likely impact of your findings such that you begin venturing prematurely into the conclusions and the recommendations. Devote all the precious space you have to your finding and your finding alone.

Characterizing the condition is the hardest job you will have. You must state the synthesized information in a way that suggests why we would want to know about it. Doing so often requires vocabulary well beyond the technical, and it requires an appreciation for the context of the report.

Is the glass half full? Or half empty? Is the program highly *fragmented*? Or just *differentiated*? Do programs work at *cross-purposes*? Are you finding a lack of *coordination*? Or are you finding a lack of *collaboration*? Coordination and collaboration are not the same thing.

We find a lot of *gaps*, but what is their nature? Is there a *displacement*? Or is there a *misalignment* of purpose or function? Don't shortchange yourself on this work. The real gift we can offer readers is shedding light on what is otherwise murky. That's the point of public policy writing—to lift the veil of obscurity on a subject and to tell stories that matter.

Conclusion: Places the report's findings in a broader context and reminds the reader of the issue's importance. In other words, the conclusion should show the reader the big picture—the significance of the findings as a whole. By showing the reader what will happen if the status quo is maintained, the conclusion also sets the stage for the recommendations you're going to make.

Again, don't get ahead of your story with statements that sound like recommendations—sentences that have a lot of *shoulds* in them. Remember that your conclusion is your chance to revisit the introduction to remind readers why they should care. It's also a chance to reestablish your criteria by noting briefly what may or may not be reasonable to expect. Given the challenge of making determinations on scarce evidence, perhaps an agency's standards are not unreasonable. You can also remind the reader what's at stake in this program or this process. Think about who your readers are, engage their interests, and appeal to their values.

Another type of perspective you can offer is one that looks at the whole as more than the sum of its parts: What is the significance of the findings taken together? The conclusion is your chance to integrate them.

The conclusion is not a recitation of your findings, though you do have to allude to them. How do you do that? It's easy—in a subordinate phrase (shown below in italics):

> *Given the lack of internal controls for program expenditures,* the program is likely to suffer continued shortfalls going forward, shortfalls that could ultimately erode support for its mission.

When you write your conclusion, pretend you are a newspaper editor, writing an editorial. Make your point strongly; this is not the place to mince words.

Recommendations: Link the root causes identified in the findings section with what needs to be done and by whom. Recommendations should flow logically from the findings and your conclusion. Keep your recommendations crisp and clear. Use active verbs so that it's clear what you want to happen. (We'll get more into using active verbs a little later on.)

Recommendations should be feasible, cost effective, and specific enough that the reader knows what we expect to happen.

Here's a checklist I use to make sure that my recommendations are well supported:

- Are there other causes that should be taken into account?
- Is the argument stated too strongly or not strongly enough based on the evidence?
- Are there circumstances in which the argument is not accurate or true?
- Could there be a solution other than the action being proposed?

Avoid Logical Fallacies

Students of public policy and those just starting off in the field often fall victim to common errors in reasoning that undermine the logic of the argument. Avoid these at all cost:

Slippery slope: This is a conclusion based on the premise that if A happens, then eventually—through a series of small steps, through B, C, . . . , X, and Y—then Z will happen, too. This type of thinking basically equates A and Z: If we don't want Z to occur, then A must not be allowed to occur either.

> *Example:* If the Department of Veterans Affairs doesn't make telemental health available to all veterans, every veteran living in rural areas of the country who are struggling with mental health issues may die by their own hands before they get the help they need.

Hasty generalization: This is a conclusion based on insufficient or biased evidence. In other words, you are rushing to a conclusion before you have all the relevant facts.

> *Example:* Unless the Department of Veterans Affairs curbs the rate of veteran suicide, the military will struggle to recruit enough new people to fill the ranks of the military.

Post hoc ergo propter hoc: This is a conclusion that assumes that if A occurred after B, then B must have caused A. More often than not, A is correlated to B but not necessarily caused by B. Cause and effect is a difficult thing to prove.

> *Example:* Suicide is caused by multiple combat deployments. As we have seen in the available research, the opposite is true, even though suicide being more common after serving in a combat zone seems logical.

Circular argument: This restates the argument rather than actually proving it.

> *Example:* Veterans have a right to mental health care; therefore, they should be allowed to get mental health treatment wherever they can.

This argument is circular because its conclusion—veterans should be able to access mental health care wherever they want—is basically the same as its second premise—veterans have a right to mental health care. Anyone who would reject the argument's conclusion should also reject its second premise and, along with it, the argument as a whole.

Either/or: This is a conclusion that oversimplifies the argument by reducing it to only two sides or choices.

> *Example:* Either programs administered by the Department of Veterans Affairs will be evaluated, or the programs will be ineffective.

Ad hominem: This is an attack on a person's character rather than a criticism of his or her opinions or arguments.

> *Example:* The reason military veterans are committing suicide is because the president doesn't care enough about them. He doesn't love America, and he doesn't do enough to support the troops when they come home.

CONCLUSION

In summary, a persuasive report will start with a "hook" and an introduction that helps set the stage for the rest of the report. You should then provide the reader with some background information, though you only need to tell as much as the reader needs to know to understand your findings. Next, you'll present your findings sections—the "meat" of the report—where you will tell the reader what is happening, what should be happening, and why it's not happening that way (the condition, criteria, and cause). After you've presented your findings, you can move to the conclusion, which is where you'll tell the reader what is going to happen if nothing is done to change what is currently happening (the effect). By this point, you will have shown the reader that there is a problem, that you know what caused it, and that there is a sense of urgency and obligation to make sure things work better in the future. Only after all of that work is completed will you be able to hit the reader with the fix—your recommendation(s). Reports structured in this way are easy to navigate and persuasive.

Before we move into the nuts and bolts of writing, here are a few high-level pieces of advice I'd like to offer you:

1. It is best and most efficient to write in three separate stages: draft first, review second, and revise third. Do not mix these tasks. Separating them allows you to manage your time and handle distractions while you write, which will help you better communicate.
2. Develop an outline that tracks with your question(s). Rework it until your story is clear and persuasive.
3. Budget your time for drafting and revising so you spend most

of it on beginnings. If you get those straight, the rest will likely follow more readily.

4. Remember the reader doesn't need to know *everything*. Some of what you found will be important to include in your story; some of it will not. As you begin to consider what to include, ask yourself, "How does this piece of evidence help me tell the story?"

5. Sometimes tables and graphics can help you illustrate a point better than text can.

Above all else, write in a clear and confident voice without appearing arrogant or intolerant. Writers who make extreme or simplistic claims, or spend most of their time attacking and demeaning the other side, weaken their own claims. Avoid emotionally loaded words. And don't merely repeat conclusions. Instead, offer information, logical reasoning, and a restrained argument to lead the reader to your conclusion. Lastly, keep your audience in mind. Use the information, facts, and logic that you think will appeal to them.

In chapters 2, 3, and 4, we'll take a look at the structure of unified and coherent paragraphs, and then we'll see what it takes to write clean and concise sentences. In chapters 5 and 6, we'll consider numerous ways you can improve your writing at the individual word level. And in chapter 7, James Bennett will show us how to use graphics and other visuals to illustrate your points and tell a good story.

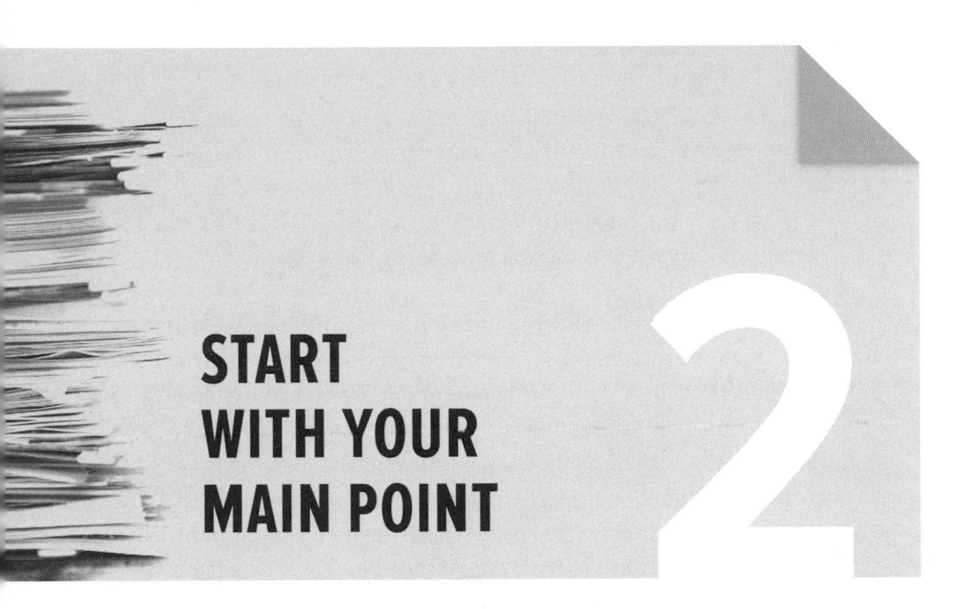

START WITH YOUR MAIN POINT

In this chapter, I show you how to grab the reader's attention right away and avoid overwhelming him or her with too much information.

One of your main goals as a public policy writer is to inform the reader—that is, explain complex phenomena—and then to persuade that reader to agree with your recommendations for improvement. The first step in achieving that goal is to structure your writing so that busy readers can easily find and understand the information they need.

Take a look at the following paragraph. Which sentence tells the main point?

> *Draft:* There is no central registry of veterans that identifies them as such, and not all veterans are counted by the Department of Veterans Affairs (VA) because only about 50 percent of veterans ever register with the VA. The Department of Defense (DOD) and the US Centers for Disease Control and Prevention have partnered with the VA to share databases in an effort to more accurately estimate the number of veterans who commit suicide, but it's still an imperfect system. The best estimate to date is that 22 veterans die by suicide daily, and more recent data suggest that the rate may be higher for veterans under the age of 30. For those still serving on active duty, DOD has a more accurate count of suicide numbers

> because it investigates all of these events. Over the last decade, the number of suicides among service members slowly increased, hitting a high of 349 in 2012 and surpassing the number of service members lost in combat. The VA cannot accurately estimate the number of veterans who commit suicide without better data.

The main point comes at the end of the paragraph: "The VA cannot accurately estimate the number of veterans who commit suicide without better data." The author of this paragraph uses what is called "inductive structure," meaning that the main point comes last—after the author traces the path he or she took in thinking through the issue.

More often than not, we write this way because it feels natural to record our thinking process, to lay out the proof for our argument. We talk this way all the time. When we tell stories or jokes, we start with all the background information and lay out a path for the listener to follow. We don't start with the moral of the story or the punch line of the joke. There's also a sort of transaction taking place when we tell stories and jokes. Your listener expects to be entertained. In public policy writing, however, there is no such expectation. That doesn't mean, of course, that we should strive to bore the reader. Instead, we should give readers what they want right away, and what they want is the main point. Only after they know the main point will they be willing to stick around and hear what else we have to say.

In public policy writing that matters, we write deductively, meaning that we present our main point first and follow it with our supporting evidence. You can use inductive or deductive reasoning to reach your conclusions, of course, but when you present those conclusions, you should deliver the conclusion first and then lay out the evidence.

Why is deductive writing better than inductive? It's simple. When our writing is deductive, our readers will better comprehend it. Take another look at the paragraph on the VA's need for better data above. Don't you feel a little lost in the beginning? It's OK if you did. If you read it again, you'll notice that it's front loaded with a series of facts and pieces of evidence—there is no central registry, only 50 percent of veterans sign up with VA, and the best estimate is that 22 veterans die by suicide daily. Without a deductive statement at the beginning to put everything into context, many readers are left wondering what it all means and why they need to know it.

Furthermore, now that most readers will be reading your writing on the Web and possibly on a device other than a desktop computer, if you don't present your main point first, your audience will likely leave before you actually get to it. Your readers simply won't have the patience required to digest an inductive report.

When we write deductively (see revised example below), our readers better understand our facts, data, and evidence because they know why they're reading about them—they want to see whether our conclusion can be supported with persuasive material.

> *Revision:* The Department of Veterans Affairs (VA) cannot accurately estimate the number of veterans who commit suicide without better data. First, there is no central registry of veterans that identifies them as such, and not all veterans are counted by the VA because only about 50 percent of veterans ever register with the VA. The Department of Defense (DOD) and the US Centers for Disease Control and Prevention have partnered with the VA to share databases in an effort to more accurately estimate the number of veterans who commit suicide, but it's still an imperfect system. The best estimate to date is that 22 veterans die by suicide daily, and more recent data suggest that the rate may be higher for veterans under the age of 30. For those still serving on active duty, DOD has a more accurate count of suicide numbers because it investigates all of these events. Over the last decade, the number of suicides among service members slowly increased, hitting a high of 349 in 2012 and surpassing the number of service members lost in combat.

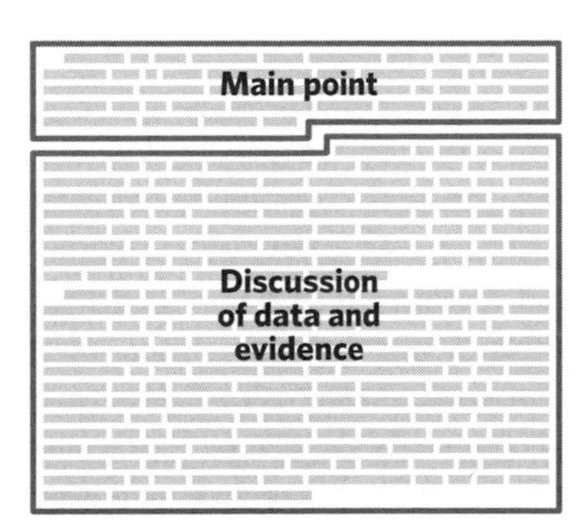

An Illustration of Deductive Structure

Writing deductively will also help you write more unified paragraphs. What does it mean to write unified paragraphs? Put simply, a unified paragraph has one main point. Do yourself a favor and commit this mantra to memory: "One paragraph, one point." If you try to pack too much into a paragraph, your readers will feel like they're sinking, and readers don't like to feel that way.

TOPIC SENTENCES

Now let's talk a bit about strong topic sentences, the most important part of a deductively written paragraph. By beginning with a strong topic sentence that tells the main point, we are able to tell the reader what the details add up to. Writing a strong topic sentence also helps you, the writer, figure out what should be included in the paragraph and, perhaps more importantly, what should be left out. Your topic sentence should answer the question: "What is the point of this paragraph?"

While this may sound easy, it is oftentimes difficult to recognize the boundaries of a main point. Writers are often forced to juggle so many overlapping ideas that it can seem impossible to deal with only one subject at a time. Readers, however, only want to try to understand one topic at a time.

In the following example, the writer discusses the methodology he employed to better understand the policies and practices the Department of Defense has in place to assist service members who are separated from the military because of a nondisability mental condition.

Take a look and consider whether the paragraph is unified. Why or why not?

> *Draft:* Over 2.6 million American men and women have deployed to Iraq and Afghanistan since September 11, 2001, and many of these individuals deployed more than once. As of March 18, 2014, nearly 52,000 have been wounded, and 5,800 have been killed in action, though these numbers do not capture the invisible costs of these wars, which include veterans we lose to suicide. The number of those who commit suicide is difficult to track because there is no central registry of veterans that identifies them as such, and not all veterans are registered with the Department of Veterans

Affairs (VA). The VA, the Department of Defense (DOD), and the US Centers for Disease Control and Prevention have partnered to share databases so that they can more accurately determine the number of veterans who commit suicide. The best estimate to date, absent strong data, is that 22 veterans die by suicide daily, though some argue that it may be as high as 30 per day. For those still serving on active duty, the DOD has a more accurate count of suicides because it investigates all of these events. Over the last decade, the number of suicides among service members slowly increased, hitting a high of 349 in 2012 and surpassing the number of service members lost in combat.

There is simply too much going on in this paragraph. While it's grammatically correct and flows nicely, it's hard to digest because it makes too many points:

Point 1: "these numbers do not capture the invisible costs of these wars, which include veterans we lose to suicide."

Point 2: "The best estimate to date, absent strong data, is that 22 veterans die by suicide daily, though some argue that it may be as high as 30 per day."

Point 3: "For those still serving on active duty, the DOD has a more accurate count of suicides because it investigates all of these events."

Now take a look at this revised version:

Revision: The total costs of the wars in Iraq and Afghanistan far surpass the numbers that are frequently quoted. Over 2.6 million American men and women have deployed to Iraq and Afghanistan since September 11, 2001, and many of these individuals have deployed more than once. In addition, as of March 18, 2014, nearly 52,000 have been wounded, and 5,800 have been killed in action. These numbers do not, however, include those we lose to suicide once they return home.

The best estimate to date, absent strong data, is that 22 veterans die by suicide daily, though some argue that it may be as high as 30 per day. The number of those who commit suicide

> is difficult to track because there is no central registry of veterans that identifies them as such, and not all veterans are registered with the Department of Veterans Affairs (VA). The VA, the Department of Defense (DOD), and the US Centers for Disease Control and Prevention have partnered to share databases so that they can more accurately determine the number of veterans who commit suicide.
>
> For those still serving on active duty, the DOD has a more accurate count of suicides because it investigates all of these events. Over the last decade, the number of suicides among service members slowly increased, hitting a high of 349 in 2012 and surpassing the number of service members lost in combat.

In the revision, the information is divided into three paragraphs, each with its own topic sentence. The revised example also has stronger topic sentences that better set the reader's expectations, something we'll discuss in the next section. Above all else, having three short paragraphs makes the information more understandable and easier for the reader to follow.

SETTING EXPECTATIONS

To help you decide what to include in a given paragraph, think about expectations and their fulfillment. A strong topic sentence will establish a reader's expectations for what will be included in that paragraph. If, after reading your topic sentence, your reader finds key terms and concepts developed in the subsequent discussion, then your writing will seem unified and logical. If, however, your reader does not find key terms and concepts developed in the subsequent discussion—or if he finds other unrelated key terms and concepts developed—then your writing will seem disorganized and illogical. Effective writers consistently use the power of the topic sentence to ensure that their paragraphs are unified.

An effective topic sentence consists of a subject and what Ostrom and Cook call a "controlling idea" (the verb and its modifiers, together called the *predicate*). The subject, in turn, should be the focus of the paragraph. The controlling idea indicates how the paragraph will support and develop the subject.

> *Sentence 1:* From fiscal years 2008 through 2012, the DOD required the four military services to monitor and report on their compliance with DOD requirements regarding separating service members with mental conditions; however, we found in the 2012 compliance report that three of the services had not achieved full compliance.
>
> *Sentence 2:* The VA system requires that veterans voluntarily participate in VA services, and not all veterans use the VA system.

After reading the sentences above, what do you expect the subsequent paragraphs to include? For sentence 1, I would expect to read about the specific requirements the DOD has implemented, as well as which services are in compliance for each requirement. Depending on how many requirements the DOD has implemented, it may be necessary to break this paragraph up with a bulleted list or other distinctive formatting. For sentence 2, I would expect to see documentation regarding the VA's policies and procedures, as well as data on how many veterans actually use the VA system. Anything not meeting these expectations will surprise your reader, and readers don't appreciate being surprised.

Once you've drafted your paragraph, a good technique for checking its unity is to (1) look for key words or phrases that appear in the topic sentence and (2) see if they appear in the remaining sentences of the paragraph. If the key words or phrases from the topic sentence do not appear in the remaining sentences, those sentences probably do not relate. If information doesn't fit, then you have to decide if it deserves a new paragraph or if it should be cut.

Remember, too, that not all topic sentences are created equal. Some are so broad that the writer could write practically anything in the paragraph. A broad topic sentence means that the controlling idea has not been adequately articulated.

> *Draft:* Military veterans face many challenges when they return home from war.
>
> *Revision:* Some of the most pressing challenges veterans face when they return home include finding meaningful employment, re-establishing connections with friends and family, and alcohol abuse.

Sometimes you may need more than one sentence to introduce your main point. If you're writing about something complex, don't be afraid to take two sentences to introduce key terms and concepts. One thing to keep in mind though: Your reader likely will not base his or her expectations on the beginning of your first topic sentence. Therefore, if you are going to use more than one sentence, you must put your most important terms and concepts at the end of the second sentence.

Moreover, topic sentences are at times unnecessary. For example, a topic sentence may not be needed if a paragraph continues to develop an idea that has been introduced clearly in a previous paragraph or if the paragraph appears in a narrative of events where generalizations might interrupt the flow. For example, the revised topic sentence about the top three challenges veterans face when they return home could serve as the topic sentence for three paragraphs. Or, you could break the original topic sentence into three topic sentences—one for each respective paragraph. Either approach could work.

Lastly, you do not have to make your point more than once in each paragraph. Many of us were taught to add a concluding sentence to each paragraph that reiterates the main point. This is not necessary in public policy writing. Simply state your point, provide your data and evidence, and move on.

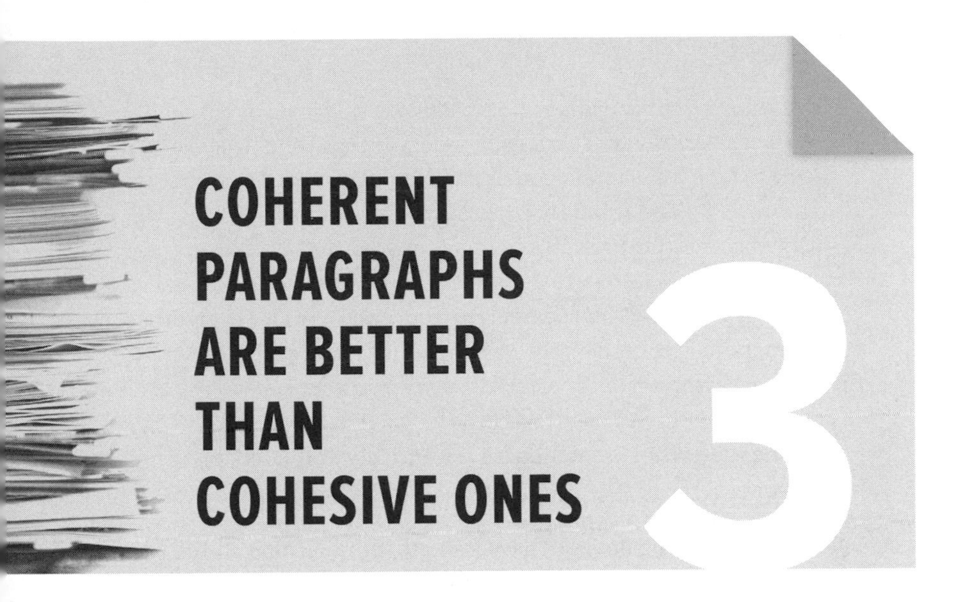

COHERENT PARAGRAPHS ARE BETTER THAN COHESIVE ONES

3

In this chapter, I show you how to ensure that only the most important information gets included in each paragraph.

Once you've mastered the art of writing deductively and limiting yourself to making one point per paragraph, you have accomplished a great deal. If, however, the sentences you write do not flow together, your reader will still judge your writing harshly.

When I talk about sentences flowing together, what I'm referring to is *coherence*. Some people confuse coherence with *cohesion*. I distinguish the two in this way: In a cohesive piece of writing, the sentences fit together one by one in the way pieces of a jigsaw puzzle do. In a coherent text, all the sentences add up to a whole, the way all the pieces in a puzzle add up to make the picture shown on the box.

This next passage has good cohesive flow because each sentence links to the next:

> My hometown, Rhinelander, Wisconsin, is home to the legendary woodland creature, the Hodag. The Hodag is a folkloric animal that is said to have the head of a frog, the face of a giant elephant, thick short legs, huge claws, the back of a dinosaur, and a long tail with spikes at the end. If you're lucky, you might catch a glimpse of the Hodag as you hike, bike, fish, paddle, shop, dine, and relax in Rhinelander. Relaxation is defined as the act of releasing tension

> and returning to equilibrium. If something is said to be out of equilibrium, it means that it's out of balance, which is something you need to have if you're going to be good at gymnastics. My high school in Rhinelander had a gymnastics team, and our mascot was, of course, the Hodag.

Though we easily can see how the sentences link to each other (the paragraph is cohesive), the passage as a whole is incoherent for three main reasons:

1. The subjects of the sentences are entirely unrelated.
2. The sentences do not share common themes or ideas.
3. The paragraph does not have a single sentence that states what the whole passage supports or explains.

A paragraph is unified (i.e., coherent) if all of the ideas in it relate to the same point. The paragraph must flow logically, transitioning from one sentence to another. "Transition is the technique of drawing sentences together," writes Thomas Whissen in *A Way with Words,* "dovetailing them, making them overlap so that the reader's journey from one sentence to the next is not a series of jerks and lunges."

Above all else, paragraph coherence depends on making apparent to the reader the logic behind the sentence order. The four major tools for improving coherence through transitions include

- following the "old-to-new" sequence,
- repeating the key words and phrases used in the topic sentence,
- using parallel structure, and
- using clear transitional words and phrases.

LINK SENTENCES BY FOLLOWING THE "OLD-TO-NEW" SEQUENCE

The old-to-new sequence is a powerful yet simple analytical tool for ensuring paragraph coherence. With the exception of the topic sentence, you should try to begin each sentence with old information, something your reader already knows. While the information may be familiar because readers have just read it in a preceding paragraph, it may also be familiar because it's general knowledge. You can then end the sentence with new information, something the reader doesn't yet know. Readers always prefer to read what's easy before what's hard, and what's familiar and simple is easier to understand than what's new and complex. When this sequence is followed consistently, your paragraphs will be coherent. (See visual below.)

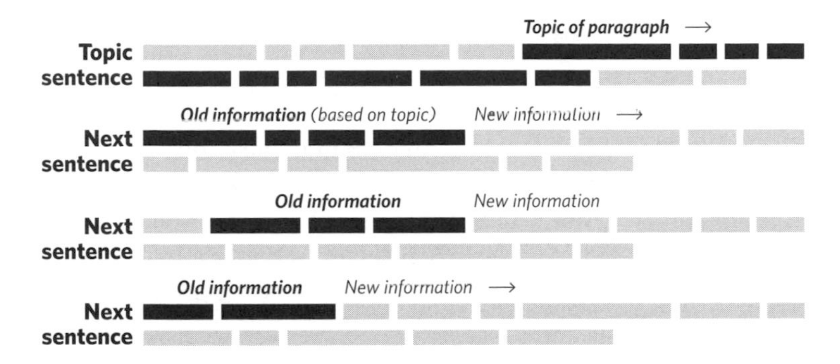

Let's take a look at this tool in action.

> Given the increased variety of support and widened reach of care available to returning veterans, their involvement in VA treatment programs is worryingly low. *This low participation rate,* coupled with an increased suicide rate, signals that many veterans are not getting the help they need from the VA's available resources. *All these resources* are opt-in, which means that veterans must actively seek them out. There are, however, many factors that prevent veterans from seeking out care. *These include,* but are not limited to, a general distrust of mental health professionals, a lack of awareness of mental health conditions, a belief that the condition is not severe enough to warrant treatment, or a belief that by seeking help they will be viewed negatively, as weak or out of control.

Readers need some of the old, simple information included in every sentence; the old information in a sentence should come *before* the new, complex information. As you can see above, the old information can be introduced in a phrase or a subordinate clause (shown above in italics), while the core of the sentence is devoted to the new information. Or, the old information might just be the repetition of a word (or its pronoun) in the new sentence.

Take this passage, for example:

> *Draft:* While all seniors age 60 and above are eligible to receive certain in-home services, most are not entitlements, which means that all seniors who would benefit from these services are not guaranteed to receive them. Congress provided approximately $1.2 billion in fiscal years 2013 and 2014 for grants to states, though in most states, this was not enough.

The sentence that addresses the amount of money Congress appropriated for this program seems out of place because it lacks any explicit relationship to the old information. Readers absolutely need that old information to orient themselves to the new information. Notice how much more clearly the passage reads after the addition of a short, orienting phrase:

> *Revision:* While all seniors age 60 and above are eligible to receive certain in-home services, most are not entitlements, which means that all seniors who would benefit from these services are not guaranteed to receive them. *To meet the needs of this vulnerable population,* Congress provided approximately $1.2 billion in fiscal years 2013 and 2014 for grants to states to provide in-home services, though in most states, this was not enough.

Note: The reference to the old information should be brief and not overshadow the new. Some writers tend to repeat far too much, thinking they are satisfying the reader. Try not to do that.

LINK SENTENCES BY REPEATING KEY WORDS AND PHRASES

Many of us have been trained to avoid repeating terms because it is judged to be boring to readers. But many readers frequently skim and scan rather than read sentence by sentence. Using the same term throughout a passage keeps the reader on track.

Repetition by careful design (as opposed to carelessness, which we'll cover later) helps readers understand that various pieces of information in a paragraph are related. Such repetition offers consistent signs to lead the reader through the paragraph. Be careful with this one, though—too much repetition or repetition of nonessential information distracts the reader and makes a paragraph dull, wordy, and redundant.

Two important uses of repetition are for accuracy and as transition:

1. *Use repetition when it is needed for accuracy.* Changing key words or phrases can confuse busy readers who could wonder whether the change denotes a difference.

> *Draft:* Under the new health care law, hospitals are required to ensure doctors comply with new rules and regulations governing outpatient care. If the hospital finds health care professionals are not complying, they are required to report those physicians and/or issue fines.

> *Revision:* Under the new health care law, hospitals are required to ensure physicians comply with new rules and regulations governing outpatient care. If the hospital finds they are not complying, the hospitals are required to report them and/or issue fines.

Notice how the variation in terms in the draft can be confusing. Is a "doctor" the same as a "health care professional"? Aren't nurses and others who administer care also considered "health care professionals"?

In the revision, we use the pronouns "they" and "them" to refer back to "physicians." The passage is much clearer as a result. In your own writing, you can also use adjective pronouns ("these problems," "such problems," or "the act"), which are useful and do not burden the reader.

2. *Use repetition to logically relate sentences.* Repeating a key word or phrase from one sentence to another helps the reader understand the relationship between the two sentences.

> The **DOD and the VA** are spending substantial *time, money, and effort* on the management of PTSD in service members and veterans. *Those efforts* have resulted in a variety of programs and services for the prevention and diagnosis of, treatment for, rehabilitation of, and research on PTSD and its comorbidities. Nevertheless, **neither department** knows with certainty whether those many programs and services are actually successful in reducing the prevalence of PTSD in service members or veterans and in improving their lives.

LINK SENTENCES BY USING PARALLEL STRUCTURE

Parallel structure links related ideas by expressing them in similar grammatical structure. Related ideas and information should have consistent verb or noun forms. Within a paragraph, certain sentences may also exhibit parallel structure by having similar introductory phrases. Such a structure, repeated in two or more sentences within a paragraph, enables the writer to show the relationship of ideas: Both structure and content convey the meaning.

> Many different sorts of transitions can be difficult for anyone who undergoes them. *For civilians,* leaving home for the first time or getting married can be stressful, especially if these transitions take place during periods of financial uncertainty. Similarly, *for returning veterans,* the challenges of coming home from a war zone can be exacerbated by uncertainty about employment, personal relationships, substance abuse, and homelessness.

Parallel structure is especially important in bulleted lists. Can you find a problem with the following list?

> A first glance at the data behind the obvious problem of veteran suicide might warrant conventional policy actions like
> - allocating more VA funding for PTSD research,
> - conduct a nationwide public awareness campaign, or
> - supplement already tedious military training programs with additional suicide education.

The second and third bullet points need to be revised to use gerunds: "conducting a nationwide public awareness campaign" and "supplementing already tedious military training programs with additional suicide education."

Bottom line: Pick one structure, and stick to it!

LINK SENTENCES BY USING TRANSITIONAL WORDS AND PHRASES

Transitional words and phrases lead the reader from one idea to another. They enhance the coherence of your writing by signaling to the reader how a particular idea logically connects to the preceding one.

Notice how the simple addition of a few transitional phrases (shown below in italics) makes clear the connection between sentences.

> We employed a number of methodologies to check for compliance among the four military services with the DOD's new requirements regarding separating service members. *First,* to analyze the extent to which the DOD and the different military services are able to identify the number of enlisted service members who were administratively separated because of a nondisability mental condition, we reviewed various documents, including the DOD's policy on the use of codes to track specific types of separations. *Second,* we interviewed DOD and military service officials to understand the type of tracking conducted and whether they maintained data on separations for nondisability mental conditions, as well as any requirements related to tracking such separations. *In addition,* we reviewed the DOD's and the military services' separation policies to identify requirements for separating service members for nondisability mental conditions, how these requirements have evolved, and whether the requirements have been consistently applied.

A list of frequently used transitional words and phrases is given in table 3.1. Be careful with transitions, though. Some writers, unfortunately, try to fake coherence by lacing their prose with conjunctions like *thus, therefore, however,* and so on, regardless of whether they signal logical connections. Experienced writers, on the other hand, may use transitional words and phrases, but they depend more on the logical flow of ideas. They are especially careful not to overuse words like *in addition, and, also, moreover, another,* and so on—words that say, simply, *Here's one more thing.* Overuse of *in addition* and the like may mean that you are simply stringing together data.

Table 3.1. FREQUENTLY USED TRANSITIONAL WORDS AND PHRASES

Addition	further, also, in addition, next, furthermore, moreover, additionally
Cause and effect	thus, as a result, therefore, consequently, because, accordingly, hence, subsequently
Comparison	similarly, likewise, in the same way
Contrast	but, yet, however, nevertheless, in contrast, on the other hand, on the contrary, instead, actually
Illustration and elaboration	for example, specifically, in particular, more precisely, in fact, indeed, more specifically, namely, that is, for instance, in other words
Numerical order	first, second, third; first, then, finally
Time	after, before, next, at the same time, currently, earlier

You need a *but* or *however* when you contradict or qualify what you just said, and you can use a *therefore* or *consequently* to wind up a line of reasoning. But avoid using words like these more than a few times per page. In fact, overuse of *however* may mean that you have not thought through the point you're trying to communicate. "Don't start a sentence with 'however,'" says William Zinsser, author of *On Writing Well,* "it hangs there like a wet dishrag. And don't end with 'however'—by that time it has lost its howeverness. Put it as early as you reasonably can. Its abruptness then becomes a virtue."

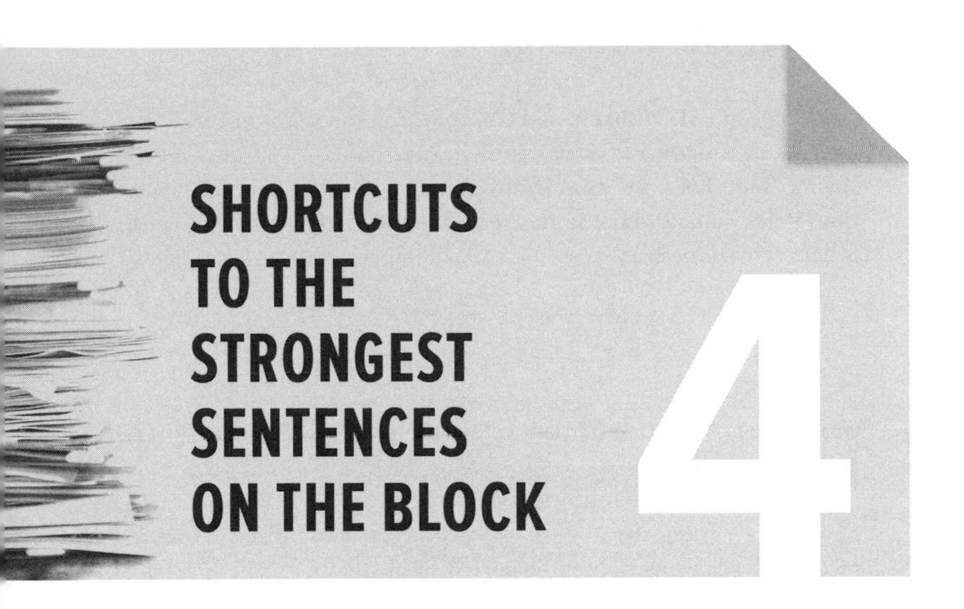

SHORTCUTS TO THE STRONGEST SENTENCES ON THE BLOCK

4

In this chapter, we move from the paragraph level to the sentence level. More specifically, I show you how to write your sentences so that your reader will be able to clearly picture what it is you're trying to say, which will help make what you're arguing much more persuasive.

As a public policy writer, you need to write sentences that are not only grammatically correct but also useful for the busy reader. Readers are on a quest for meaning. And to make meaning, your sentences need to follow certain principles. One of those principles is that every sentence needs a clear actor and a clear action.

What is the point of this sentence?

> Clinical and preventive strategies to reverse negative trends and reinforce positive trends as well as address persistent concerns are to be considered, especially if they are directed toward the veteran population as a whole with targeted messaging and intervention to each group, including young men, women, patients with and without known mental health conditions, and patients at known high risk for suicide (i.e., prior attempts).

What's the problem here? Why is it so hard to figure out the point of this sentence?

As Joseph M. Williams and Gregory Colomb point out in *Style: Lessons in Clarity and Grace,* a clear and concise sentence has a strong "sentence core" that states the (1) *actor* (the doer of the action) in the subject of the sentence and the (2) *action* (what the actor is doing) in the verb.

For example:

The Department of Veterans Affairs provided guidance.

Our actor is "The Department of Veterans Affairs." Our action is "provided guidance."

Clear and concise sentences present important information in a sentence's grammatical core—subject, verb, and object. Communication breaks down when the core is weak; that is, when it doesn't contain the important content or when its meaning is difficult for readers to grasp.

Let's take another look at the example at the beginning of this chapter. What is the actor? How about the action?

The way the sentence is currently written, the actor is "Clinical and preventive strategies to reverse negative trends and reinforce positive trends as well as address persistent concerns."

What about the action? You guessed it: "are to be considered" is the verb. Easy to follow, right? Yet the sentence still strikes us as difficult to understand.

Let's now take a closer look at four strategies we could use to improve the readability of this incredibly turgid sentence.

Place the subject close to the verb, and place the subject and verb close to the beginning of the sentence.

"Nothing more frustrates understanding," writes Richard Lauchman in *Plain Style,* than "verbs that are 'politely late' to the party. Always organize sentences so that the verb appears as close as possible to the subject." This guidance is based on the limitations of short-term memory, which we rely on when reading. If you write a sentence in which the subject is far from the verb or the verb far from the object, it may be harder for the reader to follow.

By the time readers reach the verb, they may have forgotten the subject, and by the time they reach the object, they may have forgotten the verb. Readers may have to reread the sentence when more than seven or so words come between the subject and the verb or between the verb and the object.

> *Sentence 1:* Enrollment in the program, increasing from 120,480 veterans in 1994 to 240,487 in 1998, doubled.
>
> *Sentence 2:* From 1994 to 1998, veteran enrollment in the program doubled, increasing from 120,480 to 240,487.

Why is the first sentence so much more difficult to understand than the second sentence? The answer is simple: we react differently to the two sentences because of the way the information is structured.

In sentence 1, the subject (Enrollment in the program) is followed by a 10-word phrase before we come to the verb (doubled). As readers, we're held in a state of uncertainty about what the subject is doing or what the sentence is about until we get to the verb. As a result, the reader focuses attention on the arrival of the verb and resists recognizing anything in the interrupting material as being of primary importance. Until the arrival of the verb, the reader simply sees words, not meaningful content.

In sentence 2, because the subject and verb are close together (veteran employment in the program doubled), the reader can more quickly understand the content.

Sometimes, too, we have sentences with multiple cores, and we should strive to make sure that both cores are as strong as possible.

> *Sentence 1:* The data can be parsed various ways, but suicide risk among veterans, including post-9/11 veterans, ex-service members from earlier eras, and both combat and non-combat veterans, is indisputably much greater than that of the non-veteran civilian population.
>
> *Sentence 2:* The data can be parsed various ways, but suicide risk among all veterans is indisputably much greater than that of the non-veteran civilian population. Post-9/11 vets, ex-service members from earlier eras, and both combat and non-combat veterans have higher rates of suicide.

Make sure the sentence core—subject and verb (i.e., actor and action)—contains the sentence's most important content.

An actor can be an individual, an agency or party, or even a concept—like structural functionalism or the "broken windows" theory of community policing. However, it is always better to make the actor an identifiable character rather than a concept. Readers best understand actors who actually act in the world, and they look for those actors in the subject of your sentences.

Here is a handy chart that ranks the perceptibility of different kinds of actors:

Most Perceptible → Least Perceptible:

Humans

〉Personifications (government, market)

〉Inanimate objects (receptor, fragment, instrument)

〉Familiar abstractions (assumption, distinction)

〉Specialized abstractions (polycentric governance of complex economic systems)

If an actor is not recognizable *to your readers*, then they will typically find the writing unclear.

Quite often, something can be an actor to the *writer* but not to the readers. Here is where the real trouble arises: When you use subjects that are actors to you but not to your readers, then your writing will feel clear and concise to you, but it will feel obscure and wordy to your readers.

Keeping the above in mind, let us not forget that we can use technical language. Your readers may not perceive "revenue-sharing payments from fund companies" as an actor, but that does not mean you can't use the term. Instead of using "revenue-sharing payments from fund companies" as an actor, though, put it elsewhere in your sentence. It is only in the actor position that readers are looking for actors.

If you need readers to accept an abstraction, concept, or phenomenon as an actor, then you need to devote considerable time to the task of helping them convert that new abstraction to a familiar actor. Before you set out to convert an abstraction into an actor, however, you should consider whether you really need to do so.

Here are four techniques for helping ensure that a sentence core contains the most important content:

1. *Avoid using unnecessary nominalizations.* The English language allows you to express any action with either a verb or a noun.

Who is doing what in this sentence?

> *Draft:* The Internal Revenue Service's case-by-case approach for determining fines and settlement payments could lead to inconsistent treatment of taxpayers without adequate justification for the differential treatment.

What is the Internal Revenue Service actually doing? How many times did you have to read the sentence to figure it out? Any more than once is too much!

Notice that the actor of the first sentence is 14 words long ("The Internal Revenue Service's case-by-case approach for determining fines and settlement payments"). That's a pretty weak core.

How about now?

> *Revision:* The Internal Revenue Service determines fines and settlement payments on a case-by-case basis; as a result, it may treat taxpayers inconsistently and may not be able to justify its decisions.

What's the difference between these two sentences? The first sentence is riddled with nominalizations (verbs turned into nouns). The nominalizations—approach, determining, settlement, payments, treatment (twice), and justification—make it hard to determine who is doing what. Readers will find your sentences clearer—and less bureaucratic, abstract, and difficult—if you use verbs to express important actions. Sometimes, however, we unwittingly turn a verb (determine) into a noun (determination)—a process known as "nominalization."

Nominalization might sound like jargon, but it's actually a useful term. The result of nominalization is a "smothered verb," an action that becomes lifeless. Nominalization is itself a noun derived from a verb (to nominalize). Here are some other examples:

Verb →	Nominalization	Adjective →	Nominalization
discover	discovery	careless	carelessness
move	movement	difficult	difficulty
resist	resistance	different	difference
react	reaction	elegant	elegance
fail	failure	applicable	applicability
refuse	refusal	intense	intensity

Watch for nouns (cued by suffixes like -ence, -ance, -tion, -sion, and -ment) that can serve the reader better as verbs.

Most writers of turgid prose typically use a verb not to express action but merely to state that an action exists.

> *Draft:* A need exists for greater selection efficiency.
>
> *Revision:* We must select more efficiently.

> *Draft:* An investigation of it was conducted.
>
> *Revision:* We investigated it.

> *Draft:* A review was done of the regulations.
>
> *Revision:* We reviewed the regulations.

A few patterns of useless nominalizations are easy to spot and revise. When the nominalization follows a verb with little specific meaning, change the nominalization to a verb that can replace the empty verb:

> *Draft:* The Office of Inspector General conducted an investigation into the matter.
>
> *Revision:* The Office of Inspector General investigated the matter.

Draft: The agency has no expectation that it will meet the deadline.

Revision: The agency does not expect to meet the deadline.

When the nominalization follows "there is" or "there are," change the nominalization to a verb and find a subject:

Draft: There is a need for further evaluation of this agency.

Revision: The auditor must evaluate this agency.

Draft: There was a considerable effect on the population of recipients.

Revision: The new rule considerably affected the population of recipients.

When the nominalization is the subject of an empty verb, change the nominalization to a verb and find a new subject:

Draft: The intention of the IRS is to audit the records of the program.

Revision: The IRS intends to audit the records of the program.

Draft: Our discussion concerned the implications of a cut in funding.

Revision: We discussed the implications of a cut in funding.

2. *Employ useful nominalizations.* In some cases, nominalizations are useful, even necessary. They are perfectly fine if you use them to name things; readers of English expect nouns to name things. For example, when you describe something as an "agreement," you are using a nominalization: "agreement" is nominalized from the verb "to agree." There is nothing unclear about "agreement" when you use it to name a thing, like a contract. Furthermore, nominalizations are not troublesome if you use them to name unimportant actions; they can give you a valuable tool for communicating nuances of meaning.

Nominalizations are also useful when the nominalization is a subject referring to a previous sentence. (Remember "old-to-new" information?)

> These *arguments* all depend on an incomplete data set.
>
> This *decision* could result in an unnecessary burden on recipients.

These nominalizations allow us to link sentences more cohesively.

And a nominalization can be useful when it refers to an often-repeated concept:

> Few issues have so divided Americans as *preemptive war.*
>
> The Equal Rights *Amendment* was an issue in past *elections.*

In these sentences, the nominalizations name concepts that we refer to repeatedly. Rather than spell out a familiar concept in a full clause each time we mention it, we contract it into a noun. In these cases, the abstractions often become virtual actors.

In *On Writing Well,* William Zinsser implores his readers not to "get caught holding a bag of abstract nouns," including nominalizations. "You'll sink to the bottom of the lake," he says, "and never be seen again."

When you consistently rely on verbs to express key actions, your readers benefit in multiple ways:

1. Your sentences are more concrete because they have easily recognizable actors and actions.
2. Your sentences are more concise. When you use nominalizations, you have to add articles like *a* and *the,* and prepositions such as *of, by,* and *in.* You don't need these as frequently when you use verbs and conjunctions.
3. The logic of your sentences is clearer. When you nominalize verbs, you link actions with fuzzy prepositions and phrases such as *of, by,* and *on the part of.* But when you use verbs, you link clauses with precise subordinating conjunctions such as *because, although,* and *if.*

4. Perhaps most importantly of all, your sentences tell a more coherent story, and readers will love you for that!

3. *Position attribution carefully.* In public policy writing, you will frequently need to attribute statements to sources. However, you should also think about how to best position the attribution. You can do so by considering what is more important: what was said or who said it. Notice the change in the emphasis when we revise the following sentence.

> *Draft:* An attorney from the regional solicitor's office said the primary reason why the case was settled was that some employees were uncooperative.
>
> *Revision:* The case was settled chiefly because some employees were uncooperative, according to an attorney from the regional solicitor's office.

4. *Limit the length of phrases used to connect or orient readers.* Orienters give background information. To be effective as background, orienters must give familiar information—something that the reader can quickly grasp and can use to understand the rest of the sentence. Orienters often show time or place, but they can also show purpose, condition, source, and any other information that is easy to understand quickly and that helps readers to understand the rest of the sentence.

- *Place:* In both Chicago and New York, the VA has well-established hospitals that offer inpatient care for various mental conditions.
- *Time:* From 2010 through 2012, the DOD developed criteria for evaluating its transition assistance programs.
- *Source:* According to one official we spoke to, there are many reasons why a program may choose not to collect performance data.
- *Purpose:* The trend has been to discharge service members who show signs of battle fatigue. To cope with this trend,

the VA has had to improve outreach in its warrior transition programs.

Orienters can be extremely useful to the reader—but only if you use them correctly. An orienting reference is useless if we don't know what you're referring to. Make sure your reference is specific.

Readers are looking to the beginning of the sentence for easy information. Please don't use that precious real estate to introduce complex explanations. If you need to provide such explanations as background for the next major idea, give the explanation its own sentence.

Present information in bite-sized pieces.

A sentence can be both correct and difficult to understand. The goal of a good writer is to write sentences that are both correct and clear on the first reading. This does not mean, however, that you should make the complex simplistic. Rather, it means you should make the complex *clear*. Take it from Albert Einstein: *Everything should be made as simple as possible, but not simpler.*

Readers can find a sentence needlessly challenging (unclear) because it simply exhausts their attention span. Many long sentences we write have thoughts within thoughts within thoughts. Many times, we create these types of sentences because we are trying to record the complexities of the information we are communicating. For readers, though, these sentences seem to go on and on, and may fail to resonate with them.

If you need to write a long, detailed sentence to communicate complex, nuanced information, here are two techniques that you can employ to help the reader:

1. *Break up long sentences.* The question you should ask yourself is not "how long is my sentence" but "why is this sentence so long?" If the sentence consists of a long string of prepositional phrases, it probably contains nominalizations that obscure who is doing what to whom. If, however, the sentence contains several subject/verb cores linked by connectors, then readers will be able to process it without much difficulty. Most readers of professional prose can handle long sentences just fine *if* those sentences are carefully constructed.

Improving the readability of an unwieldy sentence is sometimes as simple as breaking the sentence into two.

> *Draft:* By identifying the factors that contribute to veteran suicide—alcohol and drug abuse, depression, unemployment, and homelessness—policy makers can improve the quality of life for service members and veterans by influencing the federal government to make the necessary changes to lower veteran suicides, including increasing awareness of veteran suicides to the general public; increasing funding for the Department of Veteran Affairs to expand patient outreach and universal performance measures; and continuing long-term, national data collection efforts.

> *Revision:* By identifying the factors that contribute to veteran suicide—alcohol and drug abuse, depression, unemployment, and homelessness—policy makers can improve the quality of life for service members and veterans. More specifically, policy makers can use this knowledge to influence the federal government to make the necessary changes to lower veteran suicides, including increasing awareness of veteran suicides to the general public; increasing funding for the Department of Veteran Affairs to expand patient outreach and universal performance measures; and continuing long-term, national data collection efforts.

A shorter sentence is effective for emphasizing an important point. The longer and more complex a sentence, the harder it is for busy readers to understand its meaning. If you create too many short sentences, however, your readers may feel your writing is choppy or simplistic. Experienced writers often revise a series of short sentences into subordinate clauses or phrases, turning two or more sentences into one.

2. Carefully structure long sentences so that the central point appears first and the supporting information appears in a numbered or bulleted list. "I consider the bullet a magical device," writes Edward P. Bailey Jr., author of *The Plain English Approach to Business Writing.* It can be used "to help untangle ideas and show readers the organization within paragraphs."

Numbering

Draft: The total number of illegal immigrants in the United States at any one time is unknown because data systems among the various agencies that process these individuals are not linked so that such individuals can be readily tracked, and illegal immigrants are not assigned a unique identifier that would allow for tracking them over time.

Revision: The total number of illegal immigrants in the United States at any one time is unknown because (1) data systems among the various agencies that process these individuals are not linked so that such individuals can be readily tracked, and (2) illegal immigrants are not assigned a unique identifier that would allow for tracking them over time.

Bulleting

Draft: Since its launch in 2007, the Veterans Crisis Line has answered more than 1.6 million calls and made more than 45,000 lifesaving rescues, engaged in more than 207,700 chats, and responded to more than 32,300 texts.

Revision: Since its launch in 2007, the Veterans Crisis Line has

- answered more than 1.6 million calls,
- made more than 45,000 lifesaving rescues,
- engaged in more than 207,700 chats, and
- responded to more than 32,300 texts.

Note: Whether you use a bulleted or numbered list, create items that are parallel in structure. That means all elements that function alike must be treated alike.

Use the active voice.

Similar to avoiding unnecessary nominalizations, you can make your style more direct if you also avoid unnecessary passive voice. To apply this technique, you must first know what the active and passive voices are. A sentence is in the active voice if the subject of the sentence is the doer of the action (the actor). A sentence is in the passive voice if the object of the action is in the subject position.

Active voice: The legislator criticized the agency's position.

Passive voice: The agency's position was criticized by the legislator.

Passive voice has a terrible reputation. "Strunk and White don't speculate as to why so many writers are attracted to passive verbs," writes Stephen King in *On Writing,* "but I'm willing to; I think timid writers like them for the same reason timid lovers like passive partners. The passive voice is safe." From as early as I can remember, I've been told by English teachers that I shouldn't use passive voice—ever. This rule, however, is misleading. A passive verb is not always a problem. In fact, it can sometimes be a solution. (We'll get to that in a second.)

Active voice usually conveys information more clearly, strongly, and concisely than passive voice. "The difference," Zinsser writes, "between an active-verb style and a passive-verb style—in clarity and vigor—is the difference between life and death for a writer." Although shifting the ideas into passive voice does not change the essential meaning of the sentence, passive voice may weaken it.

A passive verb is a problem when you let it seduce you into (1) using subjects that are not actors or (2) omitting important actors altogether. Because passive voice redirects the emphasis from the actor of the action to the receiver, the actor is obscured in a prepositional phrase or left out altogether. If the actor is left out, the reader is often left with an incomplete understanding. Remember that readers of English look for the actor in the subject of the sentence. If the subject does not name an actor, the sentence immediately begins to feel unclear, indirect, and difficult.

Sometimes, though, using the passive verb can actually create a clearer sentence than using the active voice. For example, passive voice might allow you to use a subject that is an actor when using the active voice would have forced you to use a subject that is not an actor.

There are other situations in which passive voice is appropriate.

- *The actor is unknown:* The veterans' claims were misplaced, and the veterans were left on the agency's waitlist for almost two years before someone located the claims.

- *It is not important who the actor is:* The new IT infrastructure was completed early and under budget.
- *The receiver of the action, not the actor, needs to be emphasized:* Numerous veterans were sent inaccurate payments.
- *The focus needs to be kept on the same actor over two sentences (old-to-new sequence):* Veterans applying for disability benefits must first fill out an electronic application on the VA's website. In collecting evidence needed to justify their claim, veterans are encouraged by the VA to consult with their primary care providers.

In the end, use the active voice unless you have a good reason for using the passive voice.

Before we move on to talking about omitting needless words in chapter 5, let's take a quick look at an example I first found in Joseph M. Williams's *Style* on how using nominalizations and passive voice can have real-world implications.

In 1985, the Government Accounting Office (GAO)—now the Government Accountability Office—reported on why fewer than half of automobile owners who received recall letters complied with the manufacturer's instructions. The GAO found that many car owners simply could not understand the letters. The following is an example of how writers can simultaneously meet legal requirements while unwittingly (I hope, anyway!) ignoring ethical obligations. Take a look at this gem:

> A defect which involves the possible failure of a frame support plate may exist on your vehicle. This plate (front suspension pivot bar support plate) connects a portion of the front suspension to the vehicle frame, and its failure could affect vehicle directional control, particularly during heavy brake application. In addition, your vehicle may require adjustment service to the hood secondary catch system. The secondary catch may be misaligned so that the hood may not be adequately restrained to prevent hood fly-up in the event the primary latch is inadvertently left unengaged. Sudden hood fly-up beyond the secondary catch while driving could impair driver visibility. In certain cir-

> cumstances, occurrence of either of the above conditions could result in vehicle crash without prior warning.

The author of this letter nominalized all the verbs that might make a reader anxious, made most of the other verbs passive, and then deleted just about all references to the actors, particularly the manufacturer.

If you were to revise some of these sentences to make them clearer and more concise, one of the sentences, according to Williams, would surely read: "If you brake hard and the plate fails, you will not be able to steer your car." Clear writing can save lives. Simple as that.

5 WHAT NOT TO SAY

To write clearly, directly, coherently, and persuasively, we must know not only how to manage the flow of ideas but also how to express them concisely—how to "omit needless words," as Strunk and White would say. In this chapter, I show you how to ruthlessly prune your writing so that the resulting product is as long as it needs to be but as short as it can be.

In *The Elements of Style,* William Strunk Jr. wrote, "A sentence should contain no unnecessary words, a paragraph no unnecessary sentences, for the same reason that a drawing should have no unnecessary lines and a machine no unnecessary parts. This requires not that the writer make all sentences short, or avoid all detail and treat subjects only in outline, but that every word tell."

In writing public policy, however, many writers make the mistake of inflating what they write—often as a futile attempt to sound smarter or more important. As William Zinsser points out, public policy writers are not alone in this regard. "Our national tendency is to inflate," Zinsser writes. "The airline pilot who announces that he is *presently anticipating experiencing considerable precipitation* wouldn't think of saying *it may rain.* The sentence is too simple—there must be something wrong with it."

When I lived just over the border from Illinois, I commuted from Wisconsin to work in Chicago via train a few days per week.

Just before my train arrived, I would hear an announcement: "Metra commuters, your attention, please. An inbound train to Chicago is now arriving in your station. For your safety, please stand behind the yellow line until the train has come to a complete stop before boarding the train."

That announcement always bothered me because it was such a waste of breath. Allow me to explain why: *Metra commuters, your attention, please.* (No problem yet.) *An inbound train to Chicago* . . . (As opposed to an *inbound* train *from* Chicago?) *is now arriving* . . . (It is? I was wondering what the flashing lights and clanging bells were all about.) *in your station.* (I'm glad I'm not waiting for a train that's going to stop in someone else's station. By the way, is it really *my* station?) *For your safety, please stand behind the yellow line* . . . (Shouldn't I be concerned for everyone's safety? Maybe Metra is trying to appeal to my selfish inclinations?) *until the train has come to a complete stop* . . . (As opposed to a partial stop? Is there such a thing?) *before boarding the train.* (Enough people must have tried to board while the train was moving and the doors were shut that Metra felt compelled to issue this warning.)

Most of us dealt with this problem of using too many words, or trying to sound smarter than we were, while trying to impress our professors in college. It was engrained in us that longer papers were better than shorter papers and that bigger words were better than smaller ones. I once had a professor joke that the way he graded papers was by throwing them down the stairs of his house. The heavier, longer papers would make it to the bottom step. Those papers received the highest grades. The rest got correspondingly lower grades. He was kidding, of course, but he went on to tell us that if we wanted to pass his course, we would need to write until "your fingers bleed!"

And when I was in graduate school, I took great pride in using words that other people might not know or understand in context. It made me feel smarter. If someone did not understand what I wrote, it must have been because they were not as smart as I was—or at least that's what I thought then.

In the professional world we must fight this urge to sound smart. "Instead," continues Zinsser, we must "simplify, simplify. Clear thinking becomes clear writing; one can't exist without the other." Readers can more easily focus on the main idea if needless words are omitted. Our goal as public policy writers should be to eliminate words and expressions that only make our writing longer, not better.

Before I show you a number of ways of pruning your writing, I'd like to say that it is possible that you could revise the examples I use in this chapter even more radically than I did. And you would be right to do so. If in this section, however, I completely rewrote every sentence, I would show you only that it is possible to rethink the whole idea of a sentence—something that I cannot easily teach. At the end of the chapter, you would still probably not understand how I did what I did. So, for pedagogical reasons, I have stayed close to the content of each original sentence.

PRUNING NEEDLESS WORDS

Why do many of us write with needless words? First, many people begin the writing process not really understanding what they are trying to say. Many of us use the process of writing to help us think through our arguments. We are not thinking about our readers; we are struggling to get our own ideas straight. Others feel that committing ideas to the page is risky, and they worry their ideas will be attacked. As a result of this fear, they often spend too much time hiding behind introductory and often irrelevant material. Sometimes, too, we write more than we need to, and then we wait for our reviewers to tell us what actually needs to be said.

Second, many writers do not immediately know what the reader *needs* to know. We do not know what knowledge our readers have and what they will be able to infer. As a result, we often give our readers much more information than they actually need.

Lastly, most public policy writers are really busy, so they pull language from government documents or previously published reports, and they do not take the time to make sure that what they write is as clear and concise as possible.

To that end, here are 11 tips and techniques you can use to prune out unnecessary words.

1. When possible, compress several words into a word or two.

Avoid restating words and phrases that do not add meaning, and resist using excessive words that could be removed without losing the meaning of the sentence.

For example, we might be tempted to write *the reason for, for the reason that, due to the fact that, owing to the fact that, in light of the fact that, considering the fact that, on the grounds that,* or *this is why.*

Instead, we could use *because, since,* or *why.*

> *Draft:* In light of the fact that the agency received funding cuts from 2003 through 2009, it did not have enough resources to complete its projects.
>
> *Revision:* Because the agency received funding cuts . . .

Instead of writing *despite the fact that, regardless of the fact that,* or *notwithstanding the fact that,* we could use *although* or *even though.*

> *Draft:* Despite the fact that the program's performance measures were calculated several times, serious errors crept into the findings.
>
> *Revision:* Even though the program's performance measures . . .

Instead of writing *in the event that, if it should transpire/happen that,* or *under circumstances in which,* we could use *if.*

> *Draft:* In the event that the program fails to meet certain standards, its funding may be cut.
>
> *Revision:* If the program fails . . .

Instead of writing *on the occasion of, in a situation in which,* or *under circumstances in which,* we could use *when.*

> *Draft:* In a situation in which a program does not detect fraudulent applications, the program may be required to develop a more formal application procedure.
>
> *Revision:* When a program detects fraud . . .

Instead of writing *as regards, in reference to, with regard to, concerning the matter of,* or *where* x *is concerned,* we could use *about.*

Draft: The first observation I would like to make is in reference to contingency funds.

Revision: The first observation I would like to make is about contingency funds.

Instead of writing *it is crucial that, it is necessary that, there is a need/necessity for, it is important that, it is incumbent upon,* or *it cannot be avoided,* we could use *must* or *should.*

Draft: There is a need for more careful inspection of all teacher preparation programs.

Revision: We must inspect all teacher preparation programs carefully.

Instead of writing *is able to, is in a position to, has the opportunity to, has the capacity for,* or *has the ability to,* we could use *can.*

Draft: We are in a position to make a recommendation that will improve the program.

Revision: We can make a recommendation that will improve the program.

Further revision: We can recommend an improvement.

Instead of writing *it is possible that, there is a chance that, it could happen that,* or *the possibility exists for,* we could use *may.*

Draft: It is possible that nothing will come of these findings.

Revision: Nothing may come of these findings.

Instead of writing *prior to, in anticipation of, subsequent to, following on, at the same time as,* or *simultaneously with,* we could use *before.*

Draft: Prior to the expiration of the enrollment period, all forms must be submitted.

Revision: Before the enrollment period expires . . .

Some more common examples you may find helpful are shown in table 5.1.

Table 5.1. **SUBSTITUTIONS FOR WORDY PHRASES**

Wordy phrase	Consider instead
a sufficient amount of	enough
come to the conclusion	conclude
consensus of opinion	consensus
cooperated together	cooperated
despite the fact that	although
due to the fact that	because
for the purpose of	to, for
if the conditions are such that	if
in order to	to
in spite of the fact that	although
in the event that	if
is in a position to	can
located within	in
necessary requirement	required
prior to	before
subsequent to	after
utilize	use

In every class I teach at Johns Hopkins University, I show the students this table. One year, I had a student bring with him to class a sentence he had written using nearly all of the wordy and redundant phrases in the chart I had shown the class the day before. Here's what he put together:

> A sufficient amount of people have come to the conclusion that despite the fact that they are located within the same building and cooperated together in order to gain a consensus of opinion for the purpose of writing if the conditions are such that prior to attaining knowledge, they are in a position to have a necessary

requirement in the event that a lack thereof of the aforementioned knowledge is existent in spite of the fact that, and due to the fact of learning.

Not bad, huh?

"Put super simply," he told me, "it means that people come here to learn how to write better, but they don't know that yet."

2. Delete "double words."

The English language has a long tradition of doubling words—a habit we acquired shortly after we began to borrow thousands of words from Latin and French that have since been incorporated into English.

Among the common pairs are

- *full* and *complete,*
- *true* and *accurate,*
- *each* and *every,*
- *first* and *foremost,*
- *any* and *all,*
- *basic* and *fundamental,*
- *could* and *potentially,* and
- *so on* and *so forth.*

Here's the secret: Pick one. You don't need both!

3. Prune out redundant modifiers.

Delete words that are implied by other words. For example, *finish* implies *complete,* so *completely finish* is redundant. Here are a few more examples:

Basic implies *fundamental,* so *basic fundamentals* is redundant.

Important implies *essentials,* so *important essentials* is redundant.

Final implies *outcome,* so *final outcome* is redundant.

Other examples include *true facts, future plans, personal beliefs, consensus of opinion, sudden crisis, terrible tragedy, end result,* and *initial preparation.*

Here's a pretty egregious example of redundant modifiers in action:

> *Draft:* The agencies have developed a *joint, cooperative* contingency plan for *anticipating unexpected* demands on their capacity and resources before all *new upgrades* can be made to their *core, essential* systems.

Notice how much better it is when we drop those redundant modifiers:

> *Revision:* The agencies have developed a contingency plan to deal with unexpected demands on their resources before all upgrades can be made to their core systems.

4. Delete empty nouns and meaningless modifiers.

Some modifiers are what we might call "verbal tics," words that are used almost unconsciously. These can often be omitted. Words and phrases that can easily be deleted include: *kind of, really, basically, definitely, practically, actually, virtually, generally, certain, particular, individual, given, various, different, specific, for all intents and purposes.*

Take a look at this example:

> *Draft:* While this is generally seen as a logical approach that provides certain advantages, it has also practically reduced the agency's ability to control its staff's various day-to-day activities.

Look how much clearer it reads when we drop the empty nouns and meaningless modifiers:

> *Revision:* While some see this as a logical approach that provides advantages, it has also reduced the agency's ability to control its staff's day-to-day activities.

5. Avoid using adverbs as much as possible.

According to Strunk and White, adverbs can be annoying. In *On Writing Well,* William Zinsser says that "most adverbs are unnecessary" and that they "clutter" and "annoy." Stephen King cautions that "the road to hell is paved with adverbs."

Unnecessary adverbs muddy up a statement. Intensifiers—*completely, totally, absolutely*—are the most frequently overused. Used too often, these words can end up *un*intensifying a sentence that would otherwise be terse and impactful.

"Adverbs, like the passive voice, seem to have been created with the timid writer in mind," King says. "With adverbs, the writer usually tells us he or she is afraid he/she isn't expressing himself/herself clearly, that he or she is not getting the point or the picture across." King's argument isn't completely off base. Degree adverbs (used to determine the degree of an action), such as *somewhat* or *moderately,* are shunned for their tendency to make the writer seem apprehensive or fearful.

Where I tend to disagree with King, however, is about his argument regarding words like *seemingly.* While such words do convey uncertainty, they don't always convey timidity. Sometimes, uncertainty *should* be conveyed in a piece of writing. Sometimes, ambiguity is more accurate than certitude.

Degree adverbs are, of course, only one variety of this part of speech. There are plenty of others—adverbs of time, adverbs of place, and adverbs of manner—that can add clarity or complexity.

In addition, there are some verbs that cannot be fine-tuned any further. In those cases, don't be afraid to add the right adverb. The only way to learn the difference between useful and unnecessary adverbs is from the supreme writing teacher: reading. Reading great books, great magazines, great blogs—and reading a lot—allows you to internalize what works and what doesn't. Read great sentences until you can tell when one isn't. Read great paragraphs until their rhythms get stuck in your head.

Only by reading can you know when an adverb belongs in a phrase and when it belongs in the trash. In the end, when used well, adverbs serve an important purpose and can enhance writing rather than detract from it.

6. Delete words and phrases that signal you're about to offer advice or guidance.

Many writers seek to offer advice and guidance, so they add words they think will do the trick. As it turns out, however, the converse is true; the extra wording dilutes the message. Here are some examples:

Be sure to

It is important to remember that

Never forget that

There's no denying that

The truth is

The fact of the matter is

Avoid these as well:

Not surprisingly

It is common wisdom/knowledge that

Everyone recognizes/understands that

7. Avoid unnecessarily restating words and phrases.

Even the best public policy writers struggle with this one. In August 2003, the *Washington Post* published an article about the reports issued by the Government Accountability Office (GAO). The GAO, which started out as an accounting agency, has become the place Congress goes to for unbiased analysis and persuasive recommendations. The agency still has not, according to the author of the *Post* article, fully "escaped its roots in the 1950s, when its employees did, indeed, wear green eyeshades. . . . It still puts out plenty of truly turgid material. One recent 3,200 word report . . . used the phrase 'human capital' more frequently than the words 'of' or 'a.' "

As we have seen, repeating key words and phrases is a useful device we can use to let our reader know that the premise set up in the topic sentence is still being supported. We do not, however, want to unnecessarily restate words or phrases that do not add meaning or that could be removed without losing the meaning.

It is just as important to let the context establish the basis for our successive statements. If the subject is properly identified in the first sentence, it may be preferable to use a generic synonym or a pronoun in the second sentence, such as referring to the Social Security Administration as "the agency" or referring to "the former" of two already-named parties.

Let's take this paragraph as an example:

> *Draft:* When we compare the financial measures for the 17 states that index their unemployment insurance (UI) taxable wage bases with those that do not, we see the indexing states have maintained higher annual average reserve ratios and have many fewer instances of trust fund insolvency, even accounting for the smaller number of states that index. In indexing states, employers pay higher contribution rates. Benefits in indexing states relative to wages also exceed those in non-indexing states. States currently indexing their taxable wage bases also have higher trust fund reserve ratios, though 6 of the 17 indexing states have outstanding loans—as opposed to those in 25 of the 36 non-indexing states.

Now notice how much more pleasurable it is to read when, instead of unnecessarily restating words and phrases, we use pronouns:

> *Revision:* When we compare the financial measures for the 17 states that index their unemployment insurance (UI) taxable wage bases with those that do not, we see they have maintained higher annual average reserve ratios and have many fewer instances of trust fund insolvency, even accounting for their smaller numbers. In indexing states, employers pay higher contribution rates. Benefits in these states relative to wages also exceed those in the other states. States currently indexing their taxable wage bases also have higher trust fund reserve ratios, though 6 of the 17 indexing states have outstanding loans—as opposed to those in 25 of the 36 states that do not index.

Here's another example:

> *Draft:* Child Protective Services agencies collect information about children who are maltreated—including children who have died

from maltreatment—and the circumstances surrounding the maltreatment to aid efforts that prevent maltreatment.

Clearly, the writer has overused the word *maltreatment.* When this sentence was written, the writer was probably more concerned with being accurate than being concise. He or she could have simply said, "To aid in prevention efforts, Child Protective Services agencies collect information about children who are maltreated—including children who have died—and the circumstances surrounding the maltreatment." Indeed, "including children who have died," may not be accurate if the Child Protective Services only collects information on children who die as a result of maltreatment, as opposed to those who die as a result of accident or illness. A simple fix we could employ would be to fold (1) those children who are maltreated into (2) those who die as a result of their maltreatment:

> *Revision:* To aid prevention efforts, Child Protective Services agencies collect information about children who are harmed by and die from maltreatment.

Take a look at this example:

> *Draft:* The first problem we identified is excessive volatility in funding for the training program. Specifically, allocations for this training program are significantly more volatile from year to year than allocations for other programs. Some states have reported that this volatility makes program planning difficult. While some degree of volatility is to be expected due to volatility in dislocation activity, volatility in dislocations doesn't always track with volatility in funding. For several states, changes in the numbers of participants often moved in the opposite direction from changes in allocations for other programs.

Instead, we could write:

> *Revision:* The first problem we identified is excessive volatility in funding for the training program. Specifically, allocations for this training program are significantly more volatile from year to year

> than allocations for other programs, which states report results in planning difficulties. While some degree of volatility is to be expected due to dislocation activity, such activity does not track with volatility in funding. For several states, changes in the numbers of participants often moved in the opposite direction from changes in allocations for other programs.

8. Avoid redundancy in topic sentences.

We know that we need to write topic sentences that encapsulate all of the information included in a paragraph. Sometimes, however, we end up writing our paragraphs and then our topic sentences. In such cases, it's pretty easy to write topic sentences that are redundant with information included in the rest of the paragraph.

When this is the case, try using the best phrases from the sentences you have already written to craft the topic sentence.

> *Draft:* It can be challenging to detect counterfeit foreign currency. Several experts said that the wide variety of designs in which foreign currency is issued across the world makes detecting counterfeits difficult. According to one expert we talked to, there can be dozens of different versions of a single piece of currency within a single foreign country. Moreover, in some countries, outdated currency is not systematically removed from circulation. An official in one exchange office said his staff members often see multiple versions of the same currency, where security features vary from version to version, and it can be difficult to keep track of all the variations.

Which sentences could we use in the paragraph above to craft the topic sentence? What is the *actual* main point of the paragraph? When revising this paragraph, it helps to think about these three points: (1) We are trying to make the point that it is challenging to detect counterfeit foreign currency. (The original topic sentence is redundant and a bit vague.) (2) Our evidence includes testimony from several experts that it is challenging to detect counterfeit foreign currency because they come in a wide variety of designs (which is actually the main point). (3) We can use testimonial evidence from an exchange owner that corroborates what we heard from the experts.

Possible Revision: Several experts said it can be challenging to detect counterfeit foreign currency because it comes in a wide variety of designs and security features vary. According to one expert we talked to, there can be dozens of different versions of a single piece of currency within a foreign country. Moreover, in some countries, outdated currency is not systematically removed from circulation. An official in one exchange office said his staff members often see multiple versions of the same currency, where security features vary from version to version, and it can be difficult to keep track of all the variations.

9. When possible, consider combining sentences.

Wordiness also results from trying to pack too much information into one sentence or paragraph. Ironically, the attempt to be precise by using detail can have the reverse effect, resulting in sentences that sound muddled. Simple statements provide clarity, while too much detail in the wrong place can leave the reader confused. There is an art to balancing evidence with findings and detail with a broader picture. Doing so requires that we couch findings in the briefest caveats and modifiers. It also requires that we provide specifics only after we provide a context.

Rolling two sentences into one forces us to cut out the needless "filler" we often include in our first drafts.

Draft: Over the last several years, the number of children for whom states receive reimbursements has declined. Indeed, the average monthly number of children for whom states received reimbursements declined from about 200,000 in fiscal year 2007 to 169,000 in fiscal year 2013.

We could make this more concise by phrasing it this way:

Revision: Over the last several years, the average monthly number of children for whom states received reimbursements has declined from about 200,000 in fiscal year 2007 to 169,000 in fiscal year 2013.

Or, we could say,

> From fiscal year 2007 to 2013, the average monthly number of children for whom states received reimbursements declined from about 200,000 to 169,000.

Let's look at another example:

> *Draft:* Officials from the Social Security Administration said that verifying personal data—including SSNs—when Social Security benefits are claimed helps detect cases where the data are actually associated with a deceased individual. Verifying SSNs may therefore detect attempts to claim benefits fraudulently under the identity of a deceased individual.

Here is one way we could roll these two sentences together:

> *Revision:* Officials from the Social Security Administration said that verifying personal data—including SSNs—when Social Security benefits are claimed helps detect attempts to fraudulently claim benefits of deceased individuals.

The key to this technique is to first find the information that the sentences share, then delete the redundant phrase, and redraft the sentences with the best phrases rolled together.

10. Minimize the use of prepositional phrases.

Read the following sentence. Is it clear?

> *Draft:* In our interviews with officials in eight states, most could not accurately determine, based on the general definition in the law, which of their teachers met the new federal requirements.

How about now?

> *Revision:* In eight states, most officials we interviewed could not determine which of their teachers met the new federal requirements.

The first sentence uses 6 prepositional phrases and 30 words; the second uses 2 prepositional phrases and 18 words. Besides length, what's the difference? A well-structured sentence quickly directs the reader toward the main idea. A sentence with too many prepositional phrases leaves the reader feeling lost in a sea of relations, with no clear main idea.

Here are three ways to minimize the use of prepositions:

1. Express the important action in the verb.
2. Use the active voice, unless you have a good reason for using the passive voice.
3. Simplify the wording.

If a sentence has more than three prepositional phrases, consider whether readers might have difficulty focusing on the main idea. If you think they might, revise the sentence.

11. Make your point, and move on.

Often, public policy writers struggle so mightily to make their point that they inadvertently and unnecessarily repeat their point. It's best to avoid that.

> *Draft:* Due to the fact that more pension plans have become insolvent, the total amount of financial assistance the agency has provided has increased markedly in recent years. Overall, for fiscal year 2014, the agency provided $90 million in total financial assistance in order to help 45 insolvent plans cover pension benefits for about 49,000 plan participants. Generally, since 2001, the number of plans needing financial assistance has steadily increased, as has the total amount of assistance the agency has provided each year, drawing down the agency's insurance program funds. Moreover, the number of plans needing the agency's help has increased significantly in recent years, from 23 plans in fiscal year 2008 to 59 plans in fiscal year 2014. Likewise, the amount of annual assistance the agency has issued to plans has increased, from about $60 million in fiscal year 2008 to about $105 million in fiscal year 2014. (*148 words in paragraph*)

In addition to using needless phrases, including "Due to the fact that," and useless transitions like "Overall," "Generally," "Moreover," and "Likewise," the author of this paragraph repeats two main points over and over, which is completely unnecessary. The two points the writer is trying to make are:

1. The total amount of financial assistance the agency has provided has increased markedly in recent years.
2. The number of plans needing financial assistance has steadily increased.

Those two points are made three separate times, respectively.

Notice what happens when we combine the two points the writer was trying to make and then present the data that prove those points:

> *Revision:* Because more pension plans have become insolvent in recent years, the total amount of financial assistance the agency has provided has increased as well. The number of plans needing the agency's financial assistance has increased from 23 plans in fiscal year 2008 to 59 in fiscal year 2014. That year, the agency provided $105 million in total financial assistance—up from $60 million in fiscal year 2008—to help 59 insolvent plans cover pension benefits for about 49,000 plan participants. (*80 words in paragraph*)

The revised paragraph is more concise and easier to understand. In addition, it no longer needs those useless transitional phrases that were being used to string together redundant points.

Now that we're experts at pruning out needless words and phrases, let's take a look at a few strategies you can use to write about complex information in a simple, concise way.

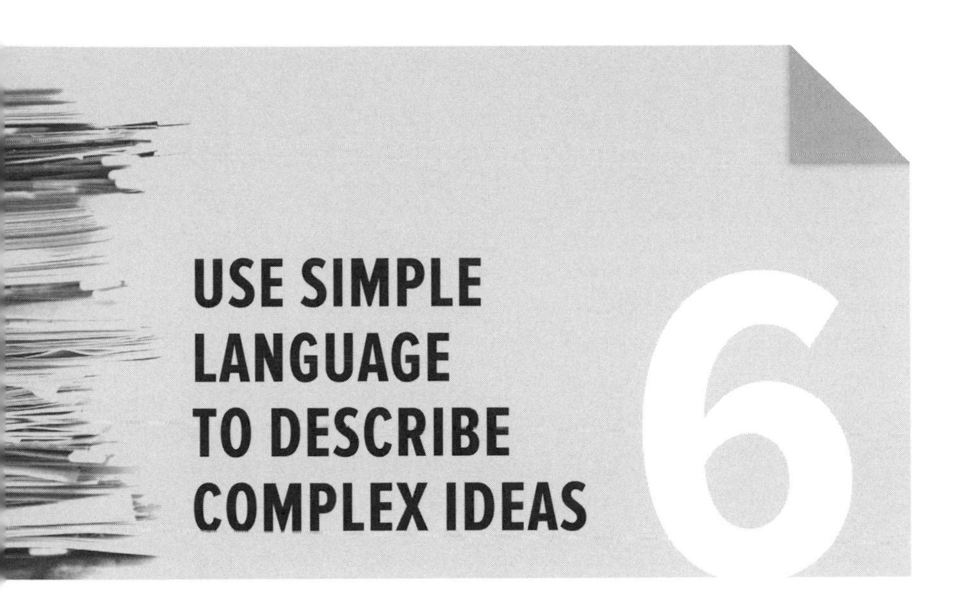

USE SIMPLE LANGUAGE TO DESCRIBE COMPLEX IDEAS

6

In this chapter, I show you how to apply a few tips and tricks that will help simplify your writing and make it more accessible to your readers.

USE COMMON WORDS

Harvard professor Steven Pinker, author of *The Sense of Style*, studies both the science of cognitive psychology—how the brain processes language, how we associate words with meanings—and the art of language. In a 2014 interview he did for the *Atlantic*, he laid out the reasons why graduate students and young professionals are often ineffective writers. When you enter graduate school, Pinker says, "Your estimate of the breadth of the knowledge of the people you are writing for gets radically miscalibrated. Highly idiosyncratic ideas are discussed as if they are common knowledge, and you lose the sense of how tiny a club you have joined." In addition, Pinker continues, "You're in terror of being judged naive and unprepared, and so you signal in your writing that you're a member of this esoteric club."

When most of us leave graduate school, we tend to keep writing for those specialists who actually understand our idiosyncratic jargon. This way of writing, however, simply won't cut it once you become a professional public policy writer. "Any specialist," writes William Zinsser in *Writing to Learn,* "who doesn't clean up his jargon when he's writing for the general population deserves to be, if not tarred, at least feathered by the mob."

Avoid meaningless jargon.

The first step to communicating complex ideas clearly is to avoid using meaningless jargon and convoluted constructions. We should only use jargon when we know our readers will be familiar with it. And even then, we should use it only when plain English will not capture the precise meaning. By choosing words that are precise and direct, we can accurately communicate with the general reader—even when our topics are complex.

> *Draft:* Since 2005, SSA has implemented expedited processing, requiring its district offices and the state disability determination offices to give priority to wounded warrior claims.
>
> *Revision:* Since 2005, SSA has expedited the processing of wounded warrior claims by directing its district offices and the state disability determination offices to give priority to these claims.

Then there are the jargon words we use to obfuscate what people are really doing. Some agencies and other institutions do this on purpose, hoping that you'll begin to use the same lexicon. Perhaps the most blatant example of this in recent memory is the language used at Guantanamo Bay Detention Camp. Since that facility opened its doors, military officials have developed their own lexicon to describe virtually everything that takes place there. For example, if a Guantanamo detainee attempts suicide, it's called "self-injurious behavior." Instead of shackles, military officials refer to the leg and wrist irons as "humane restraints." "Force-feeding" has been replaced with the clinical "enteral feeding," which replaced "assisted feeding." Even the use of "detainee"—as opposed to "prisoner"—and "detention facility" instead of "prison" was carefully thought out. In 2014, Cori Crider, the legal director of the UK-based legal organization Reprieve, which represents more than a dozen Guantanamo prisoners, told VICE News that she believes the euphemisms are an attempt to "whitewash some of the more sordid things going on, although often the effect was just to make them more sinister and Orwellian." That same year, VICE awarded Guantanamo a VICE News Award for "best use of deflective phrasing."

The list of such examples is, unfortunately, endless. Don't fall into the trap if you can avoid it!

Do not use a fancy word when a simple word will do.

Even though many a writing instructor has encouraged students to avoid overly complex words, most undergraduates, if they're being honest, will admit to deliberately using big words to appear smart. Don't do that!

According to Daniel M. Oppenheimer, a professor of psychology at Princeton University, "simpler writing is easier to process, and studies have demonstrated that processing fluency is associated with a variety of positive dimensions," including "higher judgments of truth," confidence, frequency, fame, and even "liking." In his 2006 study, titled "Consequences of Erudite Vernacular Utilized Irrespective of Necessity," Oppenheimer concludes that "overly complex texts caused readers to have negative evaluations of those texts and the associated authors, especially if the complexity was unnecessary."

There is a common word for every fancy one. When you use the common word, you rarely lose anything important. Replacing unnecessarily formal words with more common ones will make your writing sharper and more direct, and your readers will appreciate the effort.

> *Draft:* Pursuant to the recent memorandum issued November 29, 2014, because of financial exigencies, it is incumbent upon us all to endeavor to make maximal utilization of electronic communication in lieu of personal visitation.

> *Revision:* As the memo issued on November 29, 2014, said, to save money you should use email as much as you can instead of making personal visits.

Note: Writing sentences filled with needless words and phrases is, for many of us, part of the writing process. When we write, we express what we are thinking—with all the twists and turns, repetitions and hesitations. Feel free to leave this clutter in an early draft.

USE PRECISE WORDS

Before you can expect your reader to be able to understand your writing, revision is essential. Here are a few more tools we can use to make our writing more concise.

Use hedging words sparingly.

In today's world of six-second sound bites and rehearsed talking points, our institutional leaders perform strenuous verbal feats to escape having to tell us the "truth"—or even what they think. In his seminal text *On Writing Well,* William Zinsser tells us about "one classic offender" named Elliot Richardson, who held four cabinet posts in the 1970s. It's hard to choose from Richardson's trove of equivocal statements, but Zinsser asks readers to consider this one: "And yet, on balance, affirmative action has, I think, been a qualified success." There you have 5 hedging words in a 13-word sentence. And how about: "And so, at last, I come to the one firm conviction that I mentioned at the beginning: it is that the subject is too new for final judgments." This sentence reminds me of the title of a Government Accountability Office report I saw recently: "Critical Infrastructure Protection: DHS Is Taking Action, but It Is Too Early to Assess Results."

If every sentence we write expresses doubt, our writing will lack authority, and it will fail to inspire confidence and persuade. After all, the surest way to arouse and hold a reader's attention is by being specific and definite. To paraphrase Zinsser, if we bob and weave like aging boxers, we won't inspire confidence. We won't deserve it either. We can do better.

Here are some common hedging words to look out for:

- *Adverbs:* usually, often, sometimes, almost, virtually, possibly, allegedly, arguably, perhaps, apparently, in some ways, to a certain extent, somewhat, in some/certain respects

- *Adjectives:* most, many, some, a certain number of
- *Verbs:* may, might, can, could, tend, suggest, indicate

Don't get me wrong: Hedging words qualify your certainty, and they are essential to the work we do as public policy professionals. All I ask is that we use them sparingly. Save *would, should, could, may, might, can,* and other hedging words for situations involving real uncertainty.

> *Draft:* In some cases, benefit reductions can affect employers as well as plan participants. For example, representatives of one construction industry plan told us that the reduced benefits outlined in the rehabilitation plan had reduced their ability to recruit and train new apprentices. They explained that the prospect of earning only $50 of monthly retirement benefit per year of work—which after a 30-year career, would result in only $1,500 payment per month in retirement—is not very appealing to prospective employees. The officials said that, in light of the high contribution rates, this modest annuity is not enough incentive for employers to attract new employees.

> *Revision:* In some cases, benefit reductions affect not only plan participants but also employers. For example, representatives of one plan told us that the reduced benefits outlined in their rehabilitation plan had negatively affected their ability to recruit and train new apprentices. They explained that higher contribution rates would result in reduced annuity payments during retirement, which is not appealing to prospective employees.

Think about your numbers before using them.

Beware of quantity words (few, some, most, many, etc.). Such vague qualifiers could mean vastly different things to different people. For example, one person's "most" could mean over 90 percent, but for another person, it could mean anything greater than 50 percent. Here's some guidance on when to use which words:

- *Few:* less than 5 or 10 percent of the items being counted
- *Several:* more than two but less than many

- *Some:* indeterminate (You may want to avoid this one altogether.)
- *Many:* a large, undefined number
- *Majority:* 50+ percent
- *Most/Mostly:* well more than a majority but short of "almost all," (i.e., about two-thirds)
- *Generally/Largely:* more than half or more likely than not
- *Almost all:* more than 90 percent

When talking about survey results, it is acceptable to use vague qualifiers (e.g., most, few, etc.) in the heads, subheads, and executive summary. These vague qualifiers should be followed by the specific percentages in the actual text.

That's all I've got for you in terms of writing tips, tools, and tricks. In the last chapter, my colleague and friend James Bennett shows us how to use a variety of visuals to tell incredible stories. James is a recovering journalist who spent two award-winning decades bringing his visual firepower to bear at publications on both coasts and every place in between, including stints as a news artist at the *Orange County Register* and as graphics editor of the venerable *Boston Globe*.

Along the way he created and taught the first class on information graphics at the University of Missouri's famed School of Journalism. He advanced through the newspaper ranks to become assistant managing editor at the *Bakersfield Californian* before fleeing daily journalism for the glamour and excitement of federal employment. Now he helps Congress make better-informed decisions as a Visual Communication Analyst for the US Government Accountability Office. In short, he's the best at what he does.

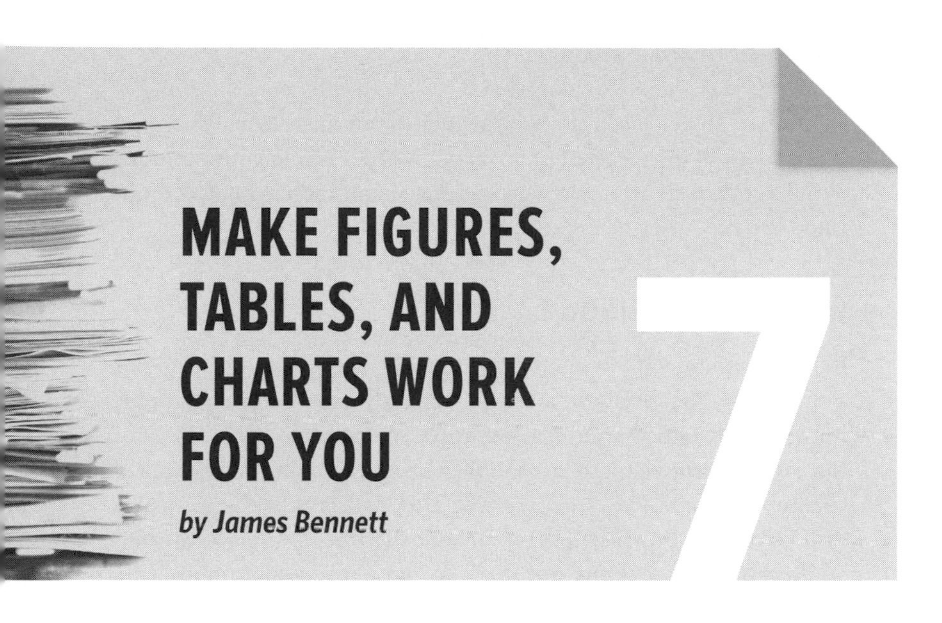

7 MAKE FIGURES, TABLES, AND CHARTS WORK FOR YOU

by James Bennett

Communicating without visuals leaves you with one very powerful hand tied behind your back. Before we take a closer look at visuals in public policy writing that matters, let's consider the following:

- The presence of a visual element makes anything much more likely to be noticed, read, and shared on social media.
- Adding visuals to a message increases the likelihood that a reader's perception of an issue will be influenced. Advertisers trade on the power of visuals!
- Numerous academic studies have shown that presenting information visually makes it dramatically easier for readers to understand and remember what you're saying.
- Scientists estimate that a sizable portion of our brains—perhaps as much as one-third—is devoted to processing visual information. That's a lot of real estate considering all the other things your brain has to do.

With the proliferation of infographics and other visuals on the Internet, it's hardly surprising that many people think graphics are a modern phenomenon. In truth, graphics have been around as long as humans have. Even the letters that form the words you're reading right now can be traced to the drawings of our ancestors. When you think about it, the rising popularity of graphics is really just a return to our roots in an attempt to make sense of an increasingly complicated and modern world.

WHAT IS A GRAPHIC?

For a discipline with ancient traditions, visual communication has a shockingly inconsistent vocabulary. Partly because every graphic is a custom act of communication, you'll find as many definitions for the term *graphic* as there are people who create them. While Shakespeare's "A rose by any other name would smell as sweet" argument is well taken, it is always helpful when we can agree on some key terms.

Practically speaking, *graphics* are an organized combination of words and images that together convey meaning and/or explain something better than either could do alone.

Although the information age has spawned a few new terms for visual communication—like *data visualization* and *infographic*—we shouldn't be dazzled by jargon. Trendy "innovations" in graphics tend to be more about format than meaningful differences. In the end, they're all just different ways of organizing and combining words and images.

There are many kinds of graphics, including charts, maps, and diagrams.

- *Charts* use math to translate numerical data into a visual form. There are many types of charts, but each type tends to be better at revealing the differences in a specific type of data set.
- *Maps* show geographic relationships and put places and distances into context, which can be critical to your reader's understanding. It's no accident that J. R. R. Tolkien put a map of Middle-earth on the opening pages of *The Hobbit*. He knew readers would struggle to keep all those places straight without a map to keep referring back to.
- *Diagrams* are like maps of an object or a system. They show the parts in the context of the whole, the players in relationship to a network, or how a system or process works. Diagrams are better than photos at explaining how something works because

the designer can remove all the unimportant details, allowing readers to focus on the point of the graphic.

- *Infographics* used to be a term for any moderately complicated visual created for a print publication. Most people now use it to describe a collection of related visual elements organized in a vertical, "scrollable" format, which makes it ideal for viewing on a computer screen. Infographics, if done well, can give an overview of a subject that may have otherwise required an extensive amount of text.

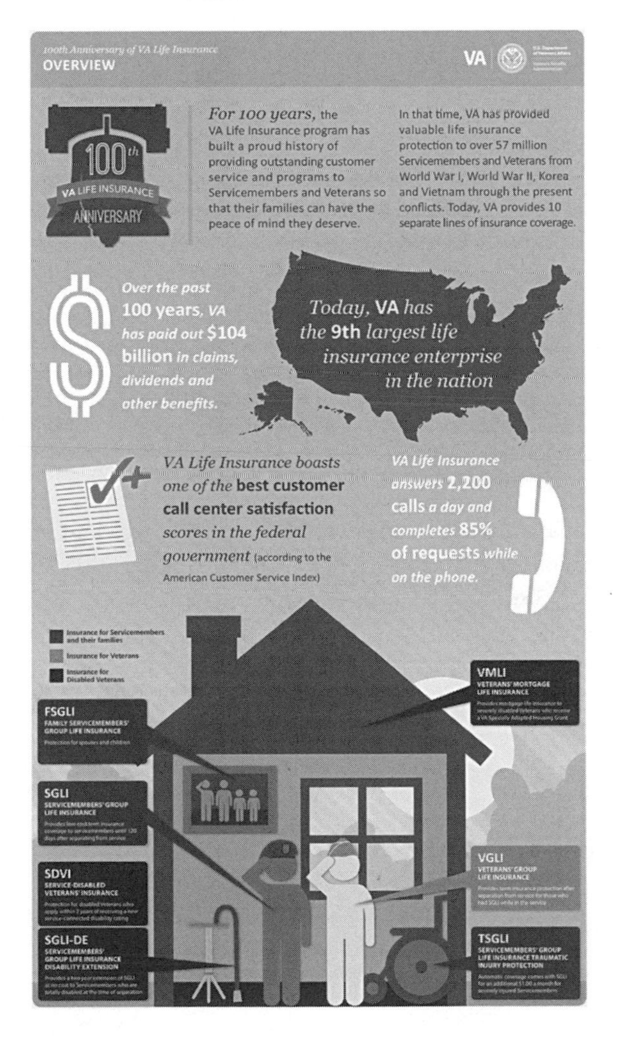

- *Data visualization* is a relatively modern term for any complex depiction of a large data set. Data visualizations tend to work better online, where users can interact with portions of the data while still seeing how each bit of information fits into a larger whole. In print, they are sometimes made up of several charts working in concert. Data visualizations are like people: They're all beautiful in their own way, but only the best have something interesting to say.

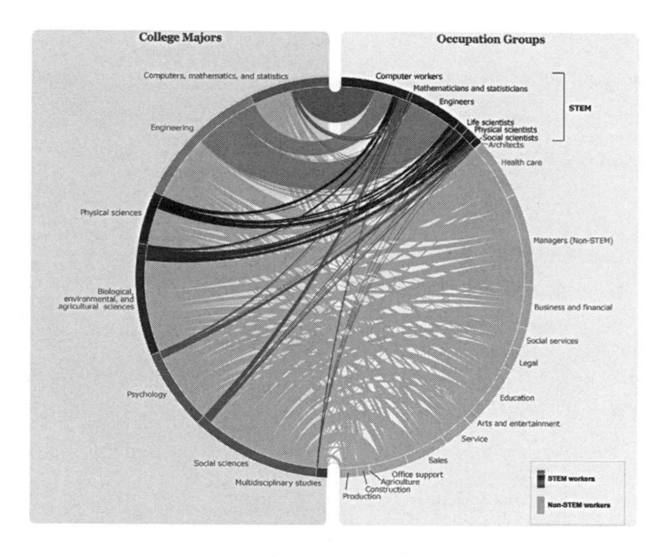

A brief note about text: The balance of words and images is an easy test of the effectiveness of a graphic. Some text is vital to letting readers in on what the visuals are trying to show, but too much text is a bad sign the visuals aren't doing enough work. For best results, limit any text to describing things the visuals can't—such as the units of a chart (i.e., "millions of dollars") or a proper name. A good rule of thumb is to try to devote no more than 20 percent of a graphic's total space to text.

THE VISUAL TOOLBOX

Experienced visual communicators treat the range of graphics like the tools on a well-stocked construction site. Things get done because they value each tool for what it does, and they take the time to match each tool to the presentation of a specific type of information. New visual communicators tend to pick the tool before fully considering the job, which is a recipe for disaster. It's not enough to know that we need to dig. The size of the required hole should dictate whether we get out the garden trowel or the backhoe.

Equipping even a simple visual toolbox can feel overwhelming, so let's start with something simple: the questioning model new journalists are taught, which was discussed in chapter 1. Reporters know they're ready to start writing when they have answers to six classic questions: Who? What? When? Where? Why? How?

Knowing which of those questions you're trying to answer visually will help suggest the type of graphic that will show it best. Once you've mastered the basic tools, you can modify and combine them as needed to create whatever will show your message best.

- *Who graphics* are called for when relationships are an important part of your story. In the public policy arena, the "who" is usually an organization, not an individual.
  - *Organizational charts* show the players and their relationships to each other using a fairly standardized "top down" format where entities (people, offices, etc.) are shown above the entities they oversee.

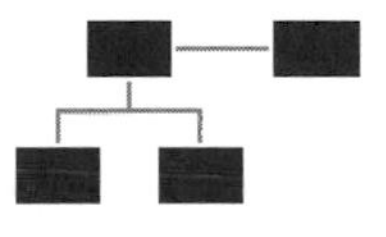

- *What graphics* are called for when you have lots of numbers or important trends to show. They're often signaled by verbs such as *rose, fell,* or *remained.* "What" questions are best answered by showing the data underlying the trend in question, and nothing does that like a chart.
  - *Charts* are all about converting numbers to a visual format, but no two types of chart present their numbers in exactly the same way. More on charts later.
  - *Photos* are another great tool for documenting current conditions because they capture *what* was happening at a particular point in time. Photos capture a tremendous amount of information, and they can be incredibly persuasive or emotional. Photos of people (especially kids)

are inherently interesting to other human beings, but many federal agencies are squeamish about identifying individuals in photographs. Know your organization's rules so that you can avoid faces or can make other adjustments when choosing what to photograph. That way, you'll end up with photos that will make it past the lawyers and into the final report.

- *When graphics* make most people think of timelines, but that's not always the best way to convey important dates to your readers.
  - *Timelines* are really a chart of how much time has elapsed between events. If the time between events isn't important, don't torture your readers with a timeline. The lines required to connect an event and its description to the right point on a timeline tend to make everything hard to read. If you really do need a timeline, keep each entry as short as possible. Five words or less is best.

    Railroads Space shuttle
    1900 1920 1940 1960 1980 2000

  - *Chronology* is the fancy name for a simple list of key events in order by date. Be sure to refer to it as a "chronology" instead of saying "let's just list the events." That tends to console anybody who's disappointed they're not getting a timeline. If they're still upset, you might point out that nobody *needs* a timeline to know that 1945 happened after 1920. A simple list of key dates is all most projects need.
- *Where graphics* are all about boundaries, geographic relationships, and scale. Most of the time, showing anything like that effectively requires a map.

— *Maps* show physical relationships, and they locate places in context to one another. A map without a scale is usually worthless because readers can't judge the distance between locations. Maps tend to be improved by adding "layers" of information, such as fire stations, areas zoned residential, or power lines—provided the additional information adds to the message.

- *Choropleth maps* assign to each geographic area (i.e., state or county) a color that varies in relation to a range of values in an underlying data set. (E.g., the darkest color shows the most activity.) Be sure to convert your data to rates (e.g., number of veterans per 10,000 residents) or percentages (e.g., percentage of draft-age adults who are veterans) when using choropleth maps. If you use raw numbers, you'll tend to "light up" the areas with higher populations because there is more activity of all kinds (veterans, skate parks, home loans) where there are more people.

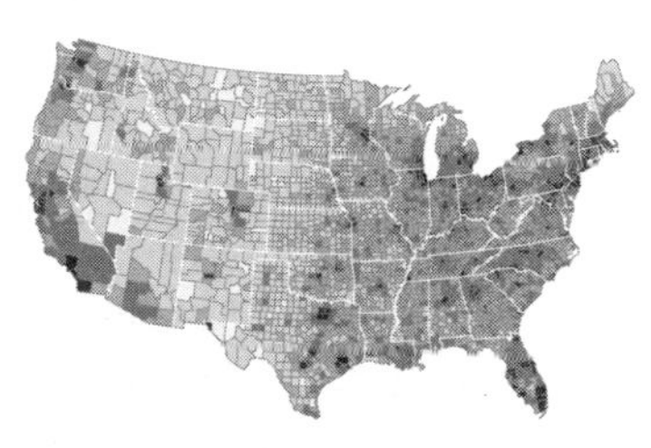

- *Data maps* combine "where" with "what" by displaying a chart element and linking it to a specific location. For example, a data map could locate every VA hospital and then put on top of that location a circle scaled to the number of veterans who used the facility last year.

- *Why graphics* are nearly impossible to do well because "why" is a nearly impossible question for graphics to answer. Nine times out of ten, you're better off not even trying. Instead, make sure you're answering "why" in the text. Getting information isn't the problem in the Internet age. The real problem is that few people have the time to sort it or explain what it means. That means readers are counting on your insights into the cause(s) of an issue. Devote as much text as possible to explaining "why" a problem is occurring. That will help inspire your newly persuaded readers to become engaged and help out.

- *How graphics* can be quite effective because "how" is the best possible question to answer with a graphic. Anytime you're faced with "how it works" or "how it happened" questions, a graphic will do a great job of telling the story. The same details that bring a paragraph to a confusing halt are the building blocks for a graphic that actually *show*s readers how something works. Think of it this way: Ikea furniture is hard enough to build with just pictures. Imagine trying to turn that pile of lumber into a bookcase with only text to guide you. That's what you're asking readers to do when you try to answer "how" questions without a graphic.
    - *Diagrams* show the parts of something and reveal how those parts work together to achieve a particular outcome. These visual depictions of physical relationships can be thought of as a kind of map of a thing or even a process. This is a step up from a regular glossary because individual components are also "defined" by what they look like and how they fit together with the other parts. But don't get overly excited about the details: Diagrams are best when you downplay, or even exclude, unimportant components.

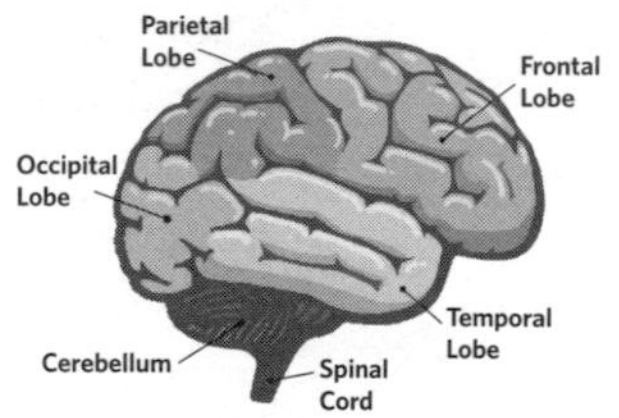

    - *Flowcharts* display the key steps in a process and what happens at critical decision points. For example, the VA enrollment process will continue to the next step when the veteran's application is complete, but it will also branch off in a new direction when an incomplete application requires staff to gather additional documentation. The conditional information in a flowchart invites readers to "play along" with a process and allows a static print graphic to feel a little like an interactive graphic on the Internet.

HOW TO TELL WHEN YOU NEED A GRAPHIC

If we tried hard enough, we could come up with a graphic for *every* element in a report. We'd have lots of visuals, but only a few of them would add anything to our message. Having too many graphics is nearly as daunting to readers as too much text. Ask yourself, "What would a graphic show?" and save your attempts at visuals for the 5Ws and an H questions that graphics answer best.

There's probably a good graphic to be done whenever you are

- *struggling with a lot of numbers in the text.* Even a relatively simple comparison gets confusing when multiple years and/or lots of numbers are involved. Replace that convoluted sentence with a chart.
- *wanting readers to remember recurring or important information.* The type of information tells you what kind of graphic you need. Places need a map; "nouns" need a diagram; dates need a chronology or timeline; a process needs a flowchart; and so on.
- *having trouble explaining how something works.* Remember the "Ikea furniture" rule, and don't risk losing readers by attempting to describe a complicated process. Show them with a diagram how the process works, and tell them in the text why they should care. Making a sketch of a process is also a great way to check and refine your own understanding of it. Start with a rough pencil sketch, and don't worry about making it pretty until you're sure readers need all the details.

FOUR WAYS TO MAKE ANY GRAPHIC BETTER

1. *Work with convention.* Cliché isn't a bad thing when it comes to visuals. Icons tend to use the "classic" representation of an item for a reason. Nobody really attends a little red schoolhouse anymore, but we still show the steeple and bell in a school icon. Remember the same principle when using color—don't use red for "cold" or put tan water around blue land on a map. Likewise, readers of English will expect to receive information in order from top to bottom, left to right. A process should flow from the first step at the left (or top) to the newest at the right (or bottom). The same goes for time in a

chart, chronology, or timeline. Start with the oldest date at the left (or top), and move to the newest at the right (or bottom). Remember, anything you do to defy convention will make it harder for readers to quickly take in the information in your graphic.

2. *Make differences mean something.* Readers are easily distracted by even small variations in a visual presentation. They will definitely notice the 12 colors you gave the bars in a chart to make it "pretty" as much (or more) than they will notice the differences in the heights of those bars. But only one of those differences is important. Make sure any variation is helping to reveal your message by keeping everything else as simple and consistent as possible. The visual representation of anything (e.g., chart elements measuring veterans, or icons showing college graduates) should remain visually consistent throughout a report.

3. *Draw the verb.* Make sure there is a visual representation of any action(s) you're describing in the text. A diagram about how veterans access VA health services should actually show a veteran going to a VA facility. A chart next to a paragraph about an increase in wait times should show the bars or lines rising, too.

4. *Adopt a visual style and stick to it.* Especially when more than one person is creating graphics, it's always better to agree ahead of time on a general look for your visuals and to share those rules with everyone. This saves your graphic creators the time of dreaming up new ways of showing things in each graphic and makes it more likely that the finished product will look like a coherent whole.

- *Use visual clues to group information.* Differences in size, style (bold, italic, etc.), and color of type help guide readers to what you want them to see. Less important information, such as axis labels, should be smaller and less visually intense than more important information, such as value labels. Color should always be used to communicate, not to decorate. Making all the water labels blue or all the veteran chart elements green helps readers group like information. Color choices should be logical, and each color should be given only one "job" per graphic. (E.g., green should not show veteran data *and* federal spending.)

- *Keep the "furniture" to a minimum.* Novice visual communicators tend to give too much emphasis to grid lines and other less important elements in charts and diagrams. Fade back or remove things like grid lines so readers can focus on the information instead of its container. Directly labeling is always better than forcing readers to "decode" the meaning of chart or map elements in a legend.

SETTING THE TABLE FOR A CHART

As we discussed earlier, charts convert numeric data into a visual format, allowing readers to easily see trends and make comparisons. No charting can begin, however, until the data are organized into the rows (horizontal) and columns (vertical) of a table. Putting data into a table also points out missing data and any trends or outliers that deserve further research. (E.g., "We're missing 2005, and the number of veterans served seems to be going down even after the 2010 reforms. I can't believe only 50 people applied in all of 2014.")

Tables and charts are both viable vehicles for information, but tables will never match the visual horsepower of charts because most readers simply won't bother with the tedious task of comparing the individual data points. Still, while tables will never be "sexy," they do have three major strengths:

1. *Tables allow readers to compare lots of unlike values.* You probably consulted a table the last time you made a major purchase to compare the key features of similar products or services. They work well because incredibly disparate measurements (e.g., dollars, weight, size, power, or time) can happily live side by side in a table. Mixing those unlike values would quickly turn a chart into an unreadable jumble.
2. *Tables convey precise values.* If the numbers you want readers to see are very close in value, and those minor differences are important, you're probably better off with a table than a chart. In most cases, readers simply can't see the differences in chart elements of nearly identical values.

3. *Tables can save pages of text.* By eliminating most of the words necessary to form complete sentences in a written-out comparison, you'll save yourself a lot of space. Plus, readers can easily find the important categories (because they're row or column labels), and none of those categories ever has to be repeated to make a comparison.

A QUICK LOOK AT CHARTS

Charts use math to translate numerical data into a visual form. This allows a reader's brain to see a pattern it can understand and remember. Entire books have been written on how to do this effectively, so we'll stick with the basics. There are many types of charts, as shown in table 7.1, but most of them work by varying *one* visual factor (e.g., length, width, height, area, etc.) per data point in an underlying data set.

Table 7.1. TYPES OF CHARTS

	Name	Best for
	Bar (aka Column)	Bar charts show one-time (e.g., daily or monthly) measurements across categories or over time. Bar charts are just as effective horizontally (with the bars pointing to the right, instead of up).
	Stacked bars	This format turns each bar into a mini pie chart. If each bar represents a different year or other unit of time, using a stacked bar will show readers how the components in each bar change over time. For best results, follow all the pie-chart rules.

Table 7.1. *(cont.)*

	Name	Best for
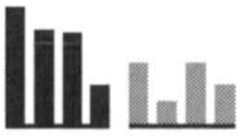	***Cluster bars***	Cluster bars "unstack" the data to let readers compare the components directly (e.g., a set of 1990 to 2000 bars for the Army, Navy, Air Force, and Marines). The message of cluster bars changes depending on their arrangement (e.g., a bar for all four armed services each year, or all the Army bars, then all the Navy bars, etc.).
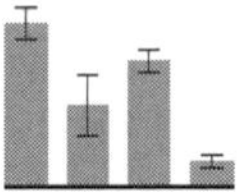	***Range charts (aka Hi-Lo or Box-and-whisker)***	They work like bars, but they add "tails" that show the lowest and highest measurement in the category each bar represents. Statisticians love the detail and accuracy in these charts, but they call a lot of attention to the outliers in each category, which may or may not be important.
	***Line (aka Fever)***	Line charts compare constant measurements across categories or over time. They tend to emphasize the trend because individual data points are hard to see on the line.
	***Pie***	Pie charts show the components of a whole by dividing up a geometric shape into segments that represent the data proportionally. It is hard for readers to discern small differences in the slices, so these charts work best for unequal divisions. Try to limit the number of slices in your pie to no more than five.
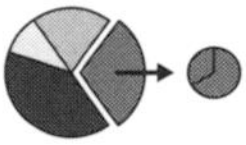	***"Zoom" pies***	When readers need to see how one slice breaks down into further slices, add a second pie chart that "zooms" out of its parent. This reveals the "X percent of Y percent" type of detail that is otherwise hard to understand. These new "sub pies" should be kept to scale (i.e., smaller) with the original pie.

(continued)

Table 7.1. *(cont.)*

	Name	Best for
	Area (aka Mountain)	If a line chart and a pie chart had a baby, it would be an area chart. These cool charts show the shift in each component's share over time. They're especially good at showing the "wave" when one category grows from a minor to a major player.
	Scatter (aka Scatter plots)	Scatter plots are the only safe way to show "apples and oranges" data in the same chart! Scatter plots work by putting one related but unalike data set on the Y axis, putting the other on the X axis, and marking their intersection with a small icon (usually a box or circle). A scatter chart with grade point average on the Y axis and total student loans on the X axis would mark the cost of each student's achievement.

CONCLUSION

There really isn't much more you need to know about writing clear and concise public policy—creating public policy writing that matters. If you can master the tips, tricks, and tools contained in this book, you'll be well on your way to changing the world.

For some people, unfortunately, being clear and concise gets confused with being unsophisticated. To them, *simple* is a dirty word. "Well, of course that story was persuasive," they argue. "It was so simple, it basically wrote itself." The fact is, however, that writing something meaningful that sounds simple is hard work. It is true that if you can learn to tell good stories as concisely as possible—and avoid getting lost in the weeds of data and analysis—then your simple prose and resulting recommendations will seem like a foregone conclusion. But that's a good thing!

Remember, too, that it takes real guts to care about something, and it takes a little hubris to stand up and offer a solution. Instead of simply complaining about the status quo, you're actually doing something about it. Do so bravely, and don't let others knock you down because your analysis seems too "simple."

On that note, let me leave you with one of my favorite movie quotes of all time. In *A League of Their Own,* manager Jimmy Dugan, played by Tom Hanks, tries to convince his star player to stick around for the first All-American Professional Girls Baseball League

Championship. He doesn't mince words about the pleasures that come with great undertakings: "It's supposed to be hard," he says. "If it wasn't hard, everyone would do it. The *hard* is what makes it great."

Good luck!

NOTES

Introduction

p. 4, "But notice the hitch here": Nassim Nicholas Taleb, *The Black Swan: The Impact of the Highly Improbable,* 2nd ed. (New York: Random House, 2010), 70.

p. 5, "Clear thinking becomes clear writing": William Zinsser, *On Writing Well: The Classic Guide to Writing Nonfiction,* 25th anniversary ed. (New York: HarperCollins, 2001), 9.

p. 6, "What we write": Joseph M. Williams and Gregory G. Colomb, *Style: Lessons in Clarity and Grace,* 10th ed. (New York: Pearson, 2010), 7.

p. 6, "In our time": George Orwell, "Politics and the English Language," in *Shooting an Elephant and Other Essays* (1950), available at www.orwell.ru/library/essays/politics/english/e_polit.

Chapter 1. Case in Point

p. 9, "According to a January 2012 study": Janet Kemp and Robert Bossarte, Department of Veterans Affairs Mental Health Services, Suicide Prevention Program, *Suicide Data Report, 2012,* www.va.gov/opa/docs/Suicide-Data-Report-2012-final.pdf.

p. 10, "We know there are a lot of suicides": Myra MacPherson, *Long Time Passing: Vietnam and the Haunted Generation* (New York: Doubleday, 1984), 252.

p. 10, "it's not only suicides": ibid., 252.

p. 11, "Veterans who die by suicide": Jan Kemp, quoted in Jeff Hargarten, Forrest Burnson, Bonnie Campo, and Chase Cook, "Suicide rate for veterans far exceeds that of civilian population," Center for Public Integrity, May 19, 2014, www.publicintegrity.org/2013/08/30/13292/suicide-rate-veterans-far-exceeds-civilian-population.

p. 11, "A 2013 long-range study": Cynthia A. LeardMann et al., "Risk Factors Associated with Suicide in Current and Former US Military Personnel," *Journal of the American Medical Association* 310 (2013): 496–506.

p. 11, "does not prove": Dr. Nancy Crum-Cianflone, quoted in Rebecca Ruiz, "Study: Deployment Not a Risk Factor for Military Suicide," *Forbes,* August 6, 2013, www.forbes.com/sites/rebeccaruiz/2013/08/06/study-deployment-not-a-risk-factor-for-military-suicide/#12fc1fcc5cf2.

p. 12, "Among other things, Kudler said": "Statement of Dr. Harold Kudler, Chief Mental Health Consultant, Veterans Health Administration (VHA), Department of Veterans Affairs (VA), Before the Committee on Veterans'

Affairs, United States Senate," November 19, 2014, www.veterans.senate.gov/imo/media/doc/VA%20Mental%20Health%20Testimony%2011.19.14.pdf.

p. 13, "With more than 22 veterans": Alex Nicholson, "IAVA Responds to Sen. Coburn's Complaints about the Clay Hunt SAV Act," *IAVA Blog,* December 12, 2014, http://iava.org/blogs/iava-responds-to-sen-coburns-blockage-of-the-clay-hunt-sav-act/.

p. 13, "After surveying": IAVA, "2014 IAVA Member Survey," 4, http://media.iava.org/IAVA_Member_Survey_2014.pdf.

p. 14, "The message I've been trying to convey": Paul Rieckhoff, quoted in T. Rees Shapiro, "Clay W. Hunt, Veterans' Advocate, Dead of Self-Inflicted Wound," *Washington Post,* April 17, 2011, www.washingtonpost.com/local/obituaries/clay-w-hunt-veterans-advocate-dead-of-self-inflicted-wound/2011/04/15/AFFkpbwD_story.html.

p. 14, "Clay enlisted": "Statement of Susan Selke, before the Senate Committee on Veterans' Affairs, for the hearing on Mental Health and Suicide Prevention," November 19, 2014, www.veterans.senate.gov/imo/media/doc/Selke%20Testimony%2011.19.14.pdf.

p. 16, "Combating veteran suicide": Rieckhoff, quoted in IAVA press release, "Clay Hunt Suicide Bill Introduced in Senate," November 17, 2014, http://iava.org/press-release/mccain-blumenthal-burr-blunt-murkowski-and-manchin-introduce-clay-hunt-suicide-prevention-bill-of-2014/.

p. 16, "I don't think this bill": Tom Coburn, quoted in Ramsey Cox, "GOP Senator Blocks Veterans' Suicide Prevention Bill," *The Hill Blog,* December 15, 2014, http://thehill.com/blogs/floor-action/senate/227223-gop-senator-blocks-veterans-suicide-prevention-bill.

p. 17, "I think some of those acts": Mark Kaplan, quoted by Jeanette Steele, "Veteran Suicides: What Might Have Saved Them," *San Diego Union-Tribune,* February 5, 2016, www.sandiegouniontribune.com/news/2016/feb/05/veterans-suicide-special-report/.

p. 19, "core American values": John J. Macionis, *Sociology,* 10th ed. (Upper Saddle River, NJ: Pearson Prentice Hall, 2005), 66.

p. 21, "generally accepted government auditing standards (GAGAS)": Comptroller General of the United States, "Government Auditing Standards: 2011 Revision," Government Accountability Office GAO-12-331G (Washington, DC: December 2011), 76, http://gao.gov/assets/590/587281.pdf.

p. 22, "laws, regulations, contracts": ibid., 6.

Chapter 3. Coherent Paragraphs Are Better Than Cohesive Ones

p. 48, "Transition": Thomas Whissen, *A Way with Words: A Guide for Writers* (New York: Oxford University Press, 1982), 111.

p. 54, "Don't start": Zinsser, *On Writing Well,* 73.

Chapter 4. Shortcuts to the Strongest Sentences on the Block

p. 56, "sentence core": Williams and Colomb, *Style,* 28.

p. 56, "Nothing more frustrates understanding," Richard Lauchman, *Plain Style: Techniques for Simple, Concise, Emphatic Business Writing* (New York: AMACOM, 1993), 65.

p. 62, "get caught": Zinsser, *On Writing Well,* 76.

p. 65, "I consider the bullet a magical device": Edward P. Bailey Jr., *The Plain English Approach to Business Writing* (New York: Oxford University Press, 1997), 77.

p. 67, "Strunk and White don't speculate": Stephen King, *On Writing: A Memoir of the Craft,* 10th anniversary ed. (New York: Pocket Books, 2000), 116.

p. 67, "The difference": Zinsser, *On Writing Well,* 67.

p. 68, "A defect": quoted in Williams and Colomb, *Style,* 59.

Chapter 5. What Not to Say

p. 71, "A sentence should contain": William Strunk Jr. and E. B. White, *The Elements of Style,* 4th ed. (Needham Heights, MA: Pearson Education, 2000), 23.

p. 71, "Our national tendency is to inflate": Zinsser, *On Writing Well,* 6.

p. 72, "Instead": ibid., 6.

p. 79, "most adverbs": ibid., 68.

p. 79, "the road to hell": King, *On Writing,* 118.

p. 79, "Adverbs": ibid., 117.

p. 80, "escaped its roots": Nicholas Thompson, "The Watchdog That's Off and Running," *Washington Post,* August 3, 2003, www.washingtonpost.com/archive/opinions/2003/08/03/the-watchdog-thats-off-and-running/de6dfb11-e21e-46cb-8261-36f359551fc3/.

Chapter 6. Use Simple Language to Describe Complex Ideas

p. 89, "Your estimate": quoted in Scott Porch, "'Literally,' Emojis, and Other Trends That Aren't Destroying English," *Atlantic,* December 30, 2014, www.theatlantic.com/entertainment/archive/2014/12/steven-pinker-interview/384092/.

p. 90, "Any specialist": William Zinsser, *Writing to Learn* (New York: Harper, 1988), 68.

p. 91, “whitewash”: quoted in Jason Leopold, “2014 VICE News Awards: Best Use of Deflective Phrasing—Guantanamo,” *VICE News,* December 22, 2014, https://news.vice.com/article/2014-vice-news-awards-best-use-of-deflective-phrasing-guantanamo.

p. 91, “simpler writing”: Daniel M. Oppenheimer, “Consequences of Erudite Vernacular Utilized Irrespective of Necessity: Problems with Using Long Words Needlessly,” *Applied Cognitive Psychology* 20 (2006): 139–56.

p. 92, “And yet, on balance”: Richardson quoted in Zinsser, *On Writing Well,* 22.

p. 92, “And so, at last”: ibid.

SUGGESTIONS FOR FURTHER READING

As I said in the introduction, my own journey as a writer, editor, and teacher has been greatly influenced by a number of fantastic books. Here is a collection I referred to continuously as I wrote this book.

Eugene Bardach. *A Practical Guide for Policy Analysis: The Eightfold Path to More Effective Problem Solving.* 5th ed. CQ Press, 2015.

Kevin Dutton. *Split-Second Persuasion: The Ancient Art and New Science of Changing Minds.* Houghton Mifflin Harcourt, 2011.

Stanley Fish. *How to Write a Sentence: And How to Read One.* HarperCollins, 1994.

Stephen King. *On Writing: A Memoir of the Craft.* 10th anniversary ed. Scribner, 2010.

Richard A. Lanham. *Revising Prose.* 5th ed. Pearson, 2006.

Richard Lauchman. *Plain Style: Techniques for Simple, Concise, Emphatic Business Writing.* AMACOM, 2008.

Donald Murray. *Writing to Deadline: The Journalist at Work.* Heinemann, 2000.

John O'Hayre. *Gobbledygook Has Gotta Go.* Nabu, 2011.

Daniel M. Oppenheimer. "Consequences of Erudite Vernacular Utilized Irrespective of Necessity." *Applied Cognitive Psychology* 20 (2006).

John Ostrom and William Cook. *Paragraph Writing Simplified.* Watson-Guptill, 1994.

Steven Pinker. *The Sense of Style: The Thinking Person's Guide to Writing in the 21st Century.* Penguin, 2015.

Nassim Nicholas Taleb. *The Black Swan: The Impact of the Highly Improbable.* 2nd ed. Random House, 2010.

William Strunk Jr. and E. B. White. *The Elements of Style.* 4th ed. Pearson, 2000.

Thomas Whissen. *A Way with Words: A Guide for Writers.* Oxford University Press, 1982.

Joseph M. Williams and Gregory G. Colomb. *Style: Lessons in Clarity and Grace.* 10th ed. Pearson, 2010.

William Zinsser. *On Writing Well: The Classic Guide to Writing Nonfiction.* 25th anniversary ed. HarperCollins, 2001.

INDEX